Medicine, Emotion and Disease, 1700–1950

Medicine, Emotion and Disease, 1700–1950

Edited by

Fay Bound Alberti

palgrave
macmillan

First published 2006 by
PALGRAVE MACMILLAN
Houndmills, Basingstoke, Hampshire RG21 6XS and
175 Fifth Avenue, New York, N.Y. 10010
Companies and representatives throughout the world

PALGRAVE MACMILLAN is the global academic imprint of the Palgrave Macmillan division of St. Martin's Press, LLC and of Palgrave Macmillan Ltd. Macmillan® is a registered trademark in the United States, United Kingdom and other countries. Palgrave is a registered trademark in the European Union and other countries.

ISBN-13: 978-1-4039-8537-8 hardback
ISBN-10: 1-4039-8537-5 hardback

This book is printed on paper suitable for recycling and made from fully managed and sustained forest sources.

A catalogue record for this book is available from the British Library.

Library of Congress Cataloging-in-Publication Data

Medicine, emotion and disease, 1700–1950 / (ed.) Fay Bound Alberti.
p. cm.
Includes bibliographical references and index.
Contents: Introduction. Emotion theory and medical history / Fay Bound Alberti — Emotions in the early modern medical tradition / Fay Bound Alberti — Patients and passions / Thomas Dixon — Languages and landscapes of emotion / Hilary Marland — "Cold calculation in the faces of horrors?" / Bertrand Taithe — Sympathy under the knife / Paul White — Fear and loathing in the laboratory and clinic / Otniel Dror — From Clever Hans to Michael Balint / Rhodri Hayward — Diagnosing with feeling / Susan Lanzoni.
ISBN 1-4039-8537-5 (cloth)
1. Emotions–Health aspects. 2. Medicine, Psychosomatic. 3. Mind and body. I. Alberti, Fay Bound, 1971-

R726.5.M437 2006
616.001'9—dc22

2006041814

10 9 8 7 6 5 4 3 2 1
15 14 13 12 11 10 09 08 07 06

Printed and bound in Great Britain by
Antony Rowe Ltd, Chippenham and Eastbourne

For Millie, Jacob and Sam

Contents

Foreword by Janet Browne ix

Acknowledgements x

Notes on Contributors xi

Introduction: Emotion Theory and Medical History xiii
Fay Bound Alberti

Chapter 1: Emotions in the Early Modern Medical Tradition 1
Fay Bound Alberti

Chapter 2: Patients and Passions: Languages of Medicine and Emotion, 1789–1850 22
Thomas Dixon

Chapter 3: Languages and Landscapes of Emotion: Motherhood and Puerperal Insanity in the Nineteenth Century 53
Hilary Marland

Chapter 4: 'Cold Calculation in the Faces of Horrors?' Pity, Compassion and the Making of Humanitarian Protocols 79
Bertrand Taithe

Chapter 5: Sympathy under the Knife: Experimentation and Emotion in Late Victorian Medicine 100
Paul White

Chapter 6: Fear and Loathing in the Laboratory and Clinic 125
Otniel E. Dror

Chapter 7: From Clever Hans to Michael Balint: Emotion, Influence and the Unconscious in British Medical Practice 144
Rhodri Hayward

Chapter 8: Diagnosing with Feeling: The Clinical Assessment of Schizophrenia in Early Twentieth-Century European Psychiatry **169**
Susan Lanzoni

Index *191*

Foreword

This volume brings together the work of eight historians specializing in the medical aspects of emotions. It began as a workshop held at the Wellcome Trust Centre for the History of Medicine at University College London in October 2004 and has grown pleasantly from there. Our objective was to explore key developments in the history of medicine and emotion in Europe and North America from the eighteenth century to the present day and to turn a fresh eye on the issues that historians face when analyzing emotions as lived experiences as well as medical categories. For 'emotions' have not only become a topic on the modern medical curriculum involving medical practitioners and patient groups in a mutual search for greater empathy and understanding, but also concern scholars investigating the history of the body, narratives of healing, the gradual politicization of the languages of feeling, the construction of gendered disease categories and the impact of our current emphasis on scientific detachment and objective diagnosis in the real world of emotional experience. However, it is clear that it is not enough merely to emote or empathize. Emotions are culturally embedded phenomena that cannot easily be captured in text and, to historians and medical professionals alike, mostly exist only as representations. All we have are representations of feelings. Much interesting work is consequently turning towards the places and 'performances' sites of these representations. The papers published here engage with the management of emotion in the patient/physician relationship from the eighteenth century, the function of gendered emotions in typically female disorders like puerperal insanity, the 'objectifications' of emotions as they began to be measured in the lab, the development of sympathy and anxiety as particularly useful categories for late-nineteenth century medical professional, and the shift from compassion to humanitarianism as a sharp political move in the present day. They are astute, innovative and lively. We are extremely grateful to the participants for their contributions and believe that their papers open important new doors to the relationships between medicine, medical practice and emotions.

Professor Janet Browne
Wellcome Trust Centre for the History of Medicine at UCL.

Acknowledgements

This collection is the product of a workshop on emotions and medicine organized by Janet Browne and myself for the Wellcome Trust in London in October 2004. I am grateful for the continued and generous financial support of the Wellcome Trust, without which this collection would not have been produced. Thanks to Janet for her insights and assistance, and to all the participants of the workshop for their valuable contributions. I am also grateful to Hal Cook, Anne Hardy and the staff at the Wellcome Trust Centre, University College London and the Wellcome Library for their encouragement and assistance during my postdoctoral research, and to all those who have commented on and discussed this work with me. I am particularly grateful to Mark Jenner and Patrick Wallis for their comments in the early stages, and to Mick Worboys, Flurin Condrau, Simon Schaffer and John Pickstone for their feedback at the publication stage. Most of all, thanks to my beloved husband, Sam Alberti, for his support, assistance and extensive reference collection. It is to him and to my beautiful children, Millie and Jacob, that this volume is dedicated.

Fay Bound Alberti
November 2005

Notes on Contributors

Fay Bound Alberti is Wellcome Research Fellow at the Centre for the History of Science, Technology and Medicine, University of Manchester. Her research interests include the history of emotions, subjectivity and illness narratives. She has previously published in journals such as the *History Workshop Journal* and *Literature and History*. Her current research project is a monograph on eighteenth- and nineteenth-century medical culture, entitled *Emotions and the Heart*.

Janet Browne lectures at the Wellcome Trust Centre for the History of Medicine at University College, London. In 2003, she completed a two-volume biography of Charles Darwin that attempted to set Darwin in scientific and personal context. She began her research life with a study of Darwin's theory of expression of the emotions. Currently she is investigating a cultural history of the gorilla, from the evolutionary debates to Dian Fossey's work in the Rwanda Mountains.

Thomas Dixon is Lecturer in History at the University of Lancaster. He is editor of the *Life and Collected Works of Thomas Brown* (Thoemmes Press, 2003), and the author of *From Passions to Emotions: The Creation of a Secular Psychological Category* (Cambridge University Press, 2003). His current research concerns the religious, scientific and political meanings of 'altruism' in Victorian Britain.

Otniel E. Dror is Head of the Section for the History of Medicine in the Medical Faculty, and Head of the Program in Ethics and Science in the Faculty of the Humanities at the Hebrew University of Jerusalem. His book is titled *Blush, Adrenaline, Excitement: Modernity and the Study of Emotions, 1860–1940*. His new book project is tentatively titled *Sensation Seeking Cultures*. Focussing on the post-Second World War period, it studies three major strands of modern life: the history of modern modes of death, the emergence of stress research and the quest for sensation/excitement in Western societies. Previous publications have appeared in *Isis*, *Configurations*, *Science in Context* and *Social Research*.

Rhodri Hayward is Research Fellow at the Exeter University Wellcome Trust Centre for Medical History. He has published on the history of religious psychology, prophecy, dreams, electrophysiology and cybernetics. His *Resisting History: Religious Transcendence and the Invention of the Unconscious* will be published by Manchester University Press in 2006. He is currently working on a short history of worry in modern Britain.

Susan Lanzoni was an NSF postdoctoral fellow at the Center for the Philosophy and History of Science, Boston University, from 2001 to 2004. She is currently a visiting assistant professor in the program for the History of Medicine and Science, Yale University. Her research interests include the history of psychiatry, experimental psychology, neurology and the sciences of mind.

Hilary Marland is Professor of History and Director of the Centre for the History of Medicine at the University of Warwick. She is the author of *Dangerous Motherhood: Insanity and Childbirth in Victorian Britain* (Palgrave Macmillan, 2004), and has published on nineteenth-century medical practice, the history of midwifery and childbirth, alternative medicine, infant welfare and women and medicine. She is also working on the politics and practices of health in the workplace, 1915–1974, and is currently developing new projects on health advice to young women, 1880–1940, and on curing with water in the English Midlands, 1840–1950.

Bertrand Taithe is Professor of Cultural History at the University of Manchester. He has published extensively on medicine and war (*Defeated Flesh*, 1999), war and citizenship (*Citizenship and Wars*, 2001) and on the history of the body. He is currently working on a major research project in the history of French humanitarianism from the 1860s to the Médecins sans Frontières and on the role of compassion in the social and cultural history of medicine.

Paul White is an editor of the *Correspondence of Charles Darwin* and lectures in the Department of History and Philosophy of Science, University of Cambridge. He is the author of *Thomas Huxley: Making the 'Man of Science'* (Cambridge University Press, 2003). He is now working on a new book project entitled *Heart and Science: Experimentation and Emotion in Late Victorian Britain*.

Introduction: Medical History and Emotion Theory

Fay Bound Alberti

The importance of emotion in healing has long been recognised by medical practitioners.[1] Yet its role in medical theory and practice is far less straightforward. Despite the rituals of distance designed to make professional medical practice unemotional as it has evolved since the eighteenth century – usually perceived as a necessary precondition for an objective diagnosis – medicine and emotions are necessarily interlinked.[2] In the present day, many physicians argue that emotional responses to illness challenge objectivity (and subsequently clinical judgement), seeking instead to engage a model of detachment. By contrast, patients complain of a lack of empathy on the part of physicians.[3] The emotional experiences of medical practitioners are also problematic, many becoming personally affected by their patients' struggles with illness. There are similar emotional difficulties for surgeons dealing with 'the possibility of failures, continuing suffering, and death on the part of their patients'.[4] Even beyond the realm of diagnosis and treatment, emotions are a live issue in the realm of medical ethics, as emotion theory has become central to medical education.[5] The notion of the 'empathic healer' or the 'empathic practitioner' has evolved at the same time, as has a range of critical writings concerned with narratives of illness and the emotional impact of physical illness on the mental wellbeing of patients.[6]

Although the question of emotional literacy and communication in modern medical practice is now problematised, its historical relationship has seldom been considered.[7] This is perhaps surprising given increased focus on the patient and the practitioner; since Roy Porter's appeal for attention to be 'paid to the objects of medicine, the recipients of the "clinical gaze", the sufferers', social historians of medicine have begun to unravel personal experiences of illness and the sociology of the medical encounter.[8] Lay perceptions of health and the relation of sick people to their physicians has been a key concern. Yet most accounts of doctor–patient relationships are concerned with patronage, power and professionalism, including such practicalities as how and why a physician might be employed, and the negotiations that went on between both parties in pursuit of a 'cure'.[9] Less attention has been paid to the emotional engagement between the patient and his or her physician, and the role of the physician's own emotions in diagnosis, prognosis and treatment. The value of such research is demonstrated by the contributions of Rhodri Hayward and Susan Lanzoni to this volume.

If the emotional context of medical practice has been understudied, there is nevertheless a growing awareness by historians of the culturally situated nature of medical diagnoses.[10] The emergence of certain medically recognised emotional pathologies, such as hysteria, hypochondria and neurasthenia, has been shown to have been socially and politically generated, revealing more about the operation of gender and class relations than any notable change in emotional life or experience.[11] Building on such 'histories from below' as stories of the insane, we also have a growing body of subjective accounts of depression, fear and melancholia.[12] Some important examples include Michael MacDonald work on the meanings of 'despair' in early modern England, which traces the etymology of such individual experiences as religious doubt, erotic melancholy and bereavement.[13] Roy Porter's analysis of Samuel Johnson's melancholy similarly presents a complex relationship between the subjective experience of sadness and the historically contingent scripts through which the emotion could be articulated and understood.[14] Accounts of the emergence of psychological categories have contributed to the understanding of the historical contingency of theories of mind, while recognisably 'modern' conditions like 'shell-shock' and 'railway spine' have become more specific targets of analyses.[15]

Categories of 'the emotional' are increasingly viewed as problematic in medical history, therefore, even the meanings of emotions themselves, and their role in the clinical encounter, have been under-explored. In seeking to redress this latter imbalance, this collection draws on social and cultural history, in particular, studies of marriage and the family, each of which has a well-established historiography dealing with emotions.[16] It also draws on recent developments in 'emotion history', which, from the groundbreaking work by Lucien Febvre in the 1940s, through a series of works by Carol and Peter Stearns in the 1980s and 1990s, has prioritised the question of emotional change, and its implications for subjectivity, interiority and selfhood.[17] Specific emotions – notably, anger, jealousy and fear – have subsequently been subject to historical scrutiny.[18] As this volume demonstrates, many of the challenges that unite these historians – including ideas about the cultural specificity of emotion, the connections between cognitive experiences and descriptions of what philosophers have called the 'raw feels' – are also shared by scholars in anthropology, sociology and psychology. [19]

For historians, the overriding theme is that of change, and the question of how far emotions can be perceived as stable entities, traceable and comparable between ages and cultures. Perhaps inevitably, therefore, much of the social historiography of emotions has been locked into debates over the quantification of emotion, past and present. While such notions as 'patriarchy', 'class' and 'gender' have been subject to scrutiny as social constructs, emotions like 'love', 'anger' and 'sadness' have been viewed as trans-historical or pre-cultural human experiences. This is most apparent in comparative analyses between early modern and modern affective standards; social and family historians have often charted the rise of sentimentalism with scant consideration of the difficulty of accessing felt emotions in the past, or of the role of emotions in social policy and institutions such as the law, or medicine.[20]

One standard position has been the identification of medieval and early modern societies as angrier and more aggressive, prone to sudden emotions and more 'childish' than our own; an evolutionary model influenced by the work of the German sociologist Norbert Elias.[21] The late seventeenth century in England is particularly associated as a turning point for emotional change; evidence for which has been found in a growth of individualistic cultural enterprises (including the rise of the novel) that have been linked to changes in individual and social patterns of affect.[22] Despite the rhetorical and structural convenience of these approaches, they fail to address the cognitive basis of emotions and the influence of culture in generating rather than suppressing or reflecting emotional experiences. Not only do they presuppose a 'hydraulic' model of emotions, which – as Barbara Rosenwein has recently argued – is itself historically contingent, they also oversimplify Elias's account of psycho- and socio-genetic transformation, so that the symbiotic relationship between self and society is overlooked.[23] They do not consider the active, shaping influences between individuals and social, medico-scientific and political institutions. They cannot, therefore, help us understand the role of emotions in doctor–patient relationships, or the gendering of particular emotional attributes at key historical moments, themes that are addressed in this volume by Rhodri Hayward and Hilary Marland, respectively.

Developments in literary theory, psychology and cultural studies have perhaps come closest to unravelling these complex interactions, through their interest in language and the production of narratives.[24] The influence of cultural construction also coalesced as an identifiable trend in emotion research in the 1980s, as seen in the work of Rom Harré, and in that of anthropologists like Catherine Lutz and Geoffrey M. White.[25] This approach allows us to see emotions as learned behaviours that owe their meaning and coherence to a series of social rules rather than innate temperaments or structures that exist outside of culture. In their purest form, constructivist approaches argue that bodies (and, by extension, emotions) do not exist outside their forms of representation and those forms are culturally situated and embedded.[26] There has been considerable reluctance by historians to plunge into the depths of pure constructivism, not least to avoid accusations of over-intellectualism. In this approach the problem of the body – and its visible physiological response mechanisms – remains.[27] For some, the solution is to discuss emotional rhetoric and display in terms of 'framing', a principle used by Charles E. Rosenberg in his discussion of the cultural construction of disease, and since taken up by other scholars.[28] Adopting the concept of framing allows us to argue in favour of the learned nature of emotional performances while retaining a flavour of lived bodily 'reality'. In this approach the emotional body continues to exist outside language. We can chart and analyse the language used to describe somatic experience, but we cannot access experience itself. This principle – of a distinction between emotion as lived and as described – is the rationale behind Carol and Peter Stearns' invention of the term 'emotionology'.[29] This label is intended to refer to 'the attitude or standards that a society, or a definable group within a society, maintains toward basic emotions and their appropriate expression', it refers to both the language employed to describe affect, and the social norms by which such affect may be measured.[30]

One characteristic of such approaches is that they continue to presuppose some essential emotional self that is influenced by, but prefigures, social life. This is also true of Rosenwein's more recent theory of 'emotional communities', and her claim that 'people move (and moved) continually from one community to another – from taverns to law courts, say – adjusting their emotional displays and their judgements of weal and woe (with greater and lesser degrees of success) to these different communities'.[31] While Rosenwein calls for histories of emotions that recognise the diverse contexts in which emotions were felt and expressed, she has yet to consider the actual processes by which the meanings of emotions themselves were produced and contested.[32]

If emotions are not merely individualistic experiences, but culturally embedded phenomena, then, how can we identify these processes of production and contestation? This difficulty becomes more acute since, as Penelope Gouk and Helen Hills have acknowledged, emotions 'cannot be accessed in unmediated form' by the historian.[33] All we have are a series of representations *about* emotion that survive in textual form. Moreover, historically there has been a vast array of conflicting discourses – religious, psychological, physiological and ideological – about emotions themselves. Since Aristotle, debates have raged about what constitutes an emotion, how many emotions are peculiar to humans and whether they originate in the mind, the soul or the body. Emotions are physical and lived experiences, giving rise to increased heartbeat, sweat, and goose bumps. Yet they are also learned and behavioural systems, revealed through gestures, posture and a series of display codes. In medieval and early modern Europe, emotions were believed to be beneficial and detrimental for the health, linked to the bodily and the secular, but also to the spiritual and divine realms.[34] They were given by God, yet (in some cases) shared with animals. They were associated with ethical principles, for good and ill. How far emotions could and should be controlled, acknowledged, denied or uncovered, has been a staple of debates over health and disease from the medieval period to the present. Crucially, moreover, as several essays in this collection demonstrate, those debates were not abstracted from, but embedded in the everyday context and construction of both social relationships and medical knowledge.

One way of discussing emotions in the past without falling into the nature/nurture debate is to focus on emotional performances in particular sites of conflict, whether that be in the courtroom, the diary or the medical encounter.[35] This means something more than asserting, as Rosenwein has, that 'people in the past, as now, expressed joy, sorrow, anger, fear, and many other feelings' in a variety of contexts. (Officials of the Angevin crown, for instance, 'participat[ed] in the culture of hatreds at home, belittling the same culture when compiling records for their Angevin masters'.)[36] For Rosenwein suggests a deliberately self-conscious crafting of pre-existing feelings; she does not examine the processes by which feelings might themselves be shaped by culture. Yet emotions themselves can be seen as concepts produced in and through social interactions. Histories of emotion could benefit from the non-Cartesian social ontology of Pierre Bourdieu in his rejection of the conventional distinctions between mental and social structures dominating Western modes of thinking about the individual.[37]

As Deborah Reed-Danahay demonstrates in *Locating Bourdieu*, the sociologist's concept of *habitus* has much relevance for the study of emotion.[38] *Habitus* encompasses 'embodied feelings and thoughts connected to commonsense understandings of the world', which are influenced by such factors as class, gender and ethnicity.[39] At its heart is the concept of embodiment, an attitude towards the internalisation and reproduction of experience which breaks down barriers between the biological and the universal in emotion theory, and by-passes debates over whether culture shapes emotions or *vice versa*. [40] Rather than viewing psychological and emotional experiences as separate from, or a product of external forces – socio-economic, political, theological or medical – Bourdieu's approach makes mental and bodily experiences of emotion a product of the internalisation of social processes, and social processes as the embodiment of individual dispositions and actions. This makes sense because emotions are, as Michelle Z. Rosaldo puts it, 'embodied thoughts'. They are not simply 'substances in our blood, but social practices organised by stories we both enact and tell. They are structured by our forms of understanding'.[41] Those 'forms of understanding' include the cultural environments in which we operate, and the institutions and structures with which we order our lives. Medicine is arguably one of the most powerful of such influences.[42]

In exploring the ways patients and physicians made sense of emotions between the eighteenth and twentieth centuries, this collection goes some way to illustrate this production of knowledge about emotion in the fields of medical theory and practice. In so doing, this volume raises many issues about emotion that have been largely unexplored by historians, though familiar to modern medical professionals. These include such questions as how and when emotions became the province of science in general and the mind in particular; the emotional nature of the relationship between medical professionals and their patients; the circumstances in which empathy and compassion were a desirable aspect of the medical encounter; and the extent to which gender might influence the diagnosis, treatment and prognosis of pathological emotional conditions. Each of the essays addresses one or more of these questions, and raises intriguing and enduring issues about the management of emotions in the therapeutic encounter, the ethics of emotion in medical research, and the relationship between *psyche* and *soma* in Western medical theory.

My account of emotions in the early modern tradition, below, not only provides historical context, but also demonstrates that there was nothing particularly inevitable about our modern perception of emotions as linked to individual subjectivities or psyches. Examining seventeenth- and eighteenth-century attitudes towards emotion prior to the emergence of laboratory medicine, it reveals the bodily nature of affect as an aspect of the mind/body/soul relation. It also outlines some of the major shifts in medico-scientific emotion theory that support historiographical appeals to the rise of modernity. One of the most important of these was the erosion of the traditional model of emotions, with the heart and vascular system at its centre, and the subsequent domination of the brain and the central nervous system. This transition was the necessary context for the scientific redefinition of emotions as measurable, quantifiable experiences manifested in and on the body – as explored by Otniel Dror – and the rise of professions uniquely equipped to deal with emotions as mental phenomena. This is not to suggest that early modern theorists of emotion did not consider the mind relevant: cognitive interpretations of the passions have long been prevalent in moral philosophy, for instance, but that conception of 'mind' incorporated the soul, not merely the brain.[43]

It is important to note that interpretations of emotions as products of mind made their logic biologically relevant, fear and anger as stimulated by internal or external events that provoke coping strategies. In the post-Freudian world we speak of 'drives' and 'impulses' as though they are the only possible interpretation of emotion and motivation – as discussed above, the hydraulic model by which we have to 'get out' emotions that are 'destructive' before we are damaged by their suppression. The language used for 'positive' or 'negative' emotions continues to be based on the evaluation of favourable or unfavourable goals: emotions drive us to approach, desire, consume and love – or reject, fear, be repulsed by and avoid – an object or person. The impetus for change is no longer rooted in the intervention of the soul, as in the early modern period, but in changes taking place in the somatic nervous system.

In many ways this functionalist approach to emotions is a development of nineteenth-century classifications, particularly Charles Darwin's *Expression of the Emotions* (1872) which expressed the evolutionary aspects of emotions as characteristics belonging to all humans and animals.[44] Like Darwin, William James stressed the biological imperative of emotions, especially in his oft-cited 1884 article 'What Is an Emotion?' although with some significant differences.[45] For James (and for the Danish physiologist Carl Lange, with whom James is often linked), the emotions occurred first in the body and then in the mind.[46] What is 'felt' at the psychological level is seen as a reaction to physiological changes occurring in the body: we feel afraid because we tremble, or distressed because we weep. With modifications this biological emphasis continues to be debated in modern psychological theories of emotion as universal principles. [47]

Yet despite the influence of Darwin and James in the production of scientific discourses of emotion, their work was less revolutionary than has often been assumed.

As Thomas Dixon and I show in this volume, there was a lengthy philosophical and medical tradition of theorising about emotion by the eighteenth century. Moreover, as Dixon demonstrates, the languages of emotions were always morally and ethically loaded. Exploring the ways in which the rhetoric of passion, affection and emotion were deployed in the medical and the political arenas, Dixon highlights differences in the meanings of the terms used by historians and by their historical subjects. While acknowledging the great diversity of theories of emotions and passions throughout the long eighteenth century, Dixon argues that a crucial transition took place in the usage of 'passions' and 'emotions' during this period, a result not only of new kinds of medical theorising, but also because of professional, institutional and disciplinary changes.[48]

Bertrand Taithe is similarly concerned with problems of the language and representation of emotional states, particularly in the context of empathy, pity and compassion. Whereas Dixon reminds us that early modern theories of affect were bound up in their moral and ethical loadings – passions being driven by the operation of the mind and the soul – Taithe argues that modern constructions of emotional response are rather more avowedly ambivalent. Taithe's consideration of humanitarian medicine, empathy and the post-nineteenth-century work of organisations such as the British Red Cross, draws attention to the importance of narrating emotions, and the complex relationship between individual experience and social acceptance. Implicitly asking broader questions about the role of emotions in institutions, Taithe considers how individual emotions became synonymous with those of a social group and, further, how compassion could be organised under the label of 'humanitarianism' in medical practice. The political nature of emotions in medical choices and availability becomes clear in Taithe's essay through his historical account of the use and abuse of pity and compassion by individuals and social groups.

The politicisation of medicine and emotion is also a key theme in Hilary Marland's consideration of motherhood and puerperal insanity in early nineteenth-century Britain. Historians have long recognised that male and female bodies were treated very differently in health and disease, and that women were believed to be prone to specific diseases, including hysteria, but few have focused specifically on discussions of the affective female body.[49] Here we see the processes of emotion management at work, but also the emergence of a newly constructed, gendered, category of emotional disorder. Marland notes that although pregnancy was associated with mental disorders prior to the nineteenth century, its appearance was deemed to be a 'natural' part of pregnancy and birth. During the nineteenth century, however, puerperal insanity was redefined as a psychiatric illness, as shifting medical paradigms helped redefine the meanings of emotional experience.

Although Marland's essay focusses on the emotions of the patient, rather than those of the physician, other contributors address the emotional content of medical practice itself. I have outlined how Taithe considers the role of emotion in constructing and maintaining medical practices. A similar theme is prominent in Paul White's discussion of medical experimentation in late Victorian medicine, and his examination of the nature of feelings and their place in laboratory practice. Part of White's account concerns the ways 'men of science' necessarily mastered their own emotions in order to convey an image of detached investigation in the scientific process. At the same time, experimenters had to explain the feelings of their animal subjects. The result, White argues, was a redefinition of the nature and meaning of feelings in the medical sciences even before emotions themselves became redefined as objects of laboratory investigation.

The role of the male physician in the clinical encounter, and this potential tension between the physician's personal and professional identities, is a subject that has attracted attention elsewhere. As Sanyantani DasGupta has recognised, illness narratives often fail to take account of the physician's investment in the therapeutic encounter, and of the feelings that a patient's condition might generate.[50] This oversight is compounded by the fact that a historical focus on patients' narratives can imply that the physician was an unemotional participant.[51] Yet the feelings of medical practitioners are central to the accounts of clinical practice given by Susan Lanzoni and Rhodri Hayward in this volume. In Lanzoni's account of the assessment of schizophrenia in early twentieth-century European psychiatry, for instance, we revisit the theme of empathy that is addressed by Taithe, but in a rather different setting. Here, Lanzoni stresses the importance of practitioners' own emotions in diagnosing and treating emotional disorders. Rather than separating their own feelings from the process of diagnosis, clinicians relied on those feelings in order to make clinical judgements. For the clinician's inability to connect or empathize with the patient indicated to the physician, on some intuitive level, that the patient suffered from schizophrenia. Moreover, despite the ideal of objectivity in most areas of the medical profession, this belief was embedded in the practice of psychiatric diagnostics well into the twentieth century.

Rhodri Hayward is similarly concerned with the active role of the practitioner in evaluating and constructing the meanings of patients' emotional experiences, and shows how the language of emotions transformed doctor–patient relations in nineteenth-century Britain. Exploring the interconnected issues of emotion, influence and the unconscious, Hayward traces concerns of 'unwitting influence' in the British response to psychoanalysis, as patients could be seen to exhibit behaviours and experiences that mirrored the expectations of their therapists. He demonstrates that the therapeutic encounter could become a 'form of theatre', in which the patients acted out a script composed by the therapist. Emotions were infectious and transferable; the condition of the patient could develop according to the emotional characteristics and propensities of the attendant physician. In this context the moral standing and personality of the physician became crucial.

Otniel Dror provides a quite different account of affect displays in his discussion of the transformation of emotions into objects of laboratory knowledge. In the life sciences, Dror observes, we encounter an alternative concept of emotion to that with which we are familiar, a category based on physical rather than mental experiences. As bodily symptoms of emotion were catalogued and evaluated, 'emotion' functioned not as a psychological category with individual meaning, but a 'visceral event', a blush or spasm that could be scientifically measured. Although the laboratory and clinic became prime sites for the production of scientific emotions, this knowledge was popularised and exported into inter-war society by a number of 'affect-gauging' technologies, such as 'Lie-Detectors' and 'Kiss-O-Meters'. At the same time as emotions were increasingly regarded as psychological events, then, their physiological significance became more relevant than ever before. The objectified body could not lie.

In each of the essays in this volume, we see the working out of medical beliefs about emotion and attempts to produce coherent scientific languages in which emotion could be understood. Medicine was not the only sphere in which emotions were contested, negotiated and produced; similar claims could be made for religious and philosophical discourses and for the range and variety of communicative practices found in everyday life. To return to Bourdieu, and to the findings of social psychology, this places emphasis on the social and constitutive nature of inner lives.[52] Rather than being distinct or separate from the social realm, concepts of 'the self' are produced and realised within it, created by an 'active relationship' between the individual and the external world.[53] Emotion can be similarly conceptualised; languages of emotion could transform doctor–patient relationships, as seen in Hayward's chapter, at the same time as new conceptions of emotional pathologies helped to define and determine the emotional experiences reported by Marland's subjects. In each case, emotion was not something interiorised, pre-existing and abstractable from its means of expression, but a performative act or concept that was realised through that expression. This approach reveals the splits and fissures, instabilities and contradictions in emotion history, and renders existing meta-narratives of change and continuity invalid.[54] Yet it also opens up new lines of enquiry, through the themes of performativity and practice, by which the historical relationship between medicine and emotion can profitably be explored.

Notes

1. For example Veronica James and Jonathan Gabe (eds), *Health and the Sociology of Emotions*, Sociology of Health and Illness Monograph Series (Cambridge, MA; Oxford: Blackwell, 1996); Bob Johnson, *Emotional Health: What Emotions Are and How They Cause Social and Mental Diseases* (York: James Naylor Foundation, 2002); Carol D. Ryff and Burton H. Singer (eds), *Emotion, Social Relationships and Health*, Series in Affective Science (Oxford: Oxford University Press, 2001); Annabelle L. Mark, *Colouring the Kaleidoscope: Emotion in Health Care Organization*, Maureen Dixon Essay Series on Health Service Organisation, 3 (London: Nuffield Trust, 2000). Keith J. Karren Brent Q. Hafen, N. Lee Smith and Kathryn J. Frandsen (eds), *Mind/Body Health: The Effects of Attitudes, Emotions, and Relationships* (San Francisco; London: Pearson/Benjamin Cummings, 2006).
2. On the necessity of detachment, see John L. Coulehan, 'Tenderness and Steadiness: Emotions in Medical Practice', *Literature and Medicine* (1995), 14, pp. 222–36.
3. For an engaging analysis of the cognitive and ethical role of emotions like empathy in medical practice, see Jodi Halpern, *From Detached Concern to Empathy: Humanizing Medical Practice* (Oxford: Oxford University Press, 2001), reviewed by Lainie Friedman Ross, *Perspectives in Biology and Medicine* (2003), 46, pp. 310–1.
4. Kirsti Torjuul, Ann Nordam and Venke Sørlie, 'Ethical Challenges in Surgery as Related by Practicing Surgeons', *BMC Medical Ethics* (2005), 6, 2, p. 2.
5. C. Leget, 'Avoiding Evasion: Medical Ethics Education and Emotion Theory', *Journal of Medical Ethics* (2004), 30, pp. 490–3.
6. Important examples include Michael J. Bennett, *The Empathic Healer: An Endangered Species?* Practical Resources for the Mental Health Professional (San Diego; London: Academic Press, 2001); Ellen Singer More and Maureen

A. Milligan (eds), *The Empathic Practitioner: Empathy, Gender and Medicine* (New Brunswick, NJ: Rutgers University Press, 1994); Howard M. Spiro Mary G. McCrea Curnen, Enid Peschel and Deborah St James (eds), *Empathy and the Practice of Medicine: Beyond Pills and the Scalpel* (New Haven, CT; London: Yale University Press, 1993); Bernard Lown, *The Lost Art of Healing: Practicing Compassion in Medicine* (Boston: Houghton Mifflin, 1996). On narratives and illness, see Arthur W. Frank, *The Wounded Storyteller: Body, Illness, and Ethics* (Chicago; London: University of Chicago Press, 1995); Arthur Kleinman, *The Illness Narratives: Suffering, Healing and the Human Condition* (New York: Basic Books, 1989); and the influential Susan Sontag, *Illness as Metaphor and: AIDS and its Metaphors* (London: Penguin, 2002).

7. For a dated though still relevant discussion, see George Rosen, 'People, Disease and Emotion: Some Newer Problems for Research in Medical History', *Bulletin of the History of Medicine* (1967), 41, pp. 5–23.
8. Roy Porter, 'Introduction' in idem (ed.), *Patients and Practitioners: Lay Perceptions of Medicine in Pre-Industrial Society*, Cambridge Studies in the History of Medicine (Cambridge: Cambridge University Press, 1985), p. 1; idem, *A Social History of Madness: Stories of the Insane* (London: Weidenfeld and Nicolson, 1987); Lilian R. Furst, *Between Doctors and Patients: The Changing Balance of Power* (Charlottesville: University Press of Virginia, 1998); Mary E. Fissell, 'The Disappearance of the Patient's Narrative and the Invention of Hospital Medicine', in Roger French and Andrew Wear (eds), *British Medicine in an Age of Reform,* Wellcome Institute Series in the History of Medicine (London; New York: Routledge, 1991), pp. 91–109; Edward Shorter, *Bedside Manners: The Troubled History of Doctors and Patients* (New York : Simon and Schuster, 1985).
9. N. Jewson, 'The Disappearance of the Sick Man from Medical Cosmology, 1770–1870', *Sociology* (1976), 10, pp. 225–244; idem, 'Medical Knowledge and the Patronage System in Eighteenth-Century England', *Sociology* (1974), 8, pp. 369–85; Margaret Pelling, 'Barbers and Barber-Surgeons: An Occupational Group in an English Provincial Town, 1550–1640', *The Society for the Social History of Medicine Bulletin* (1981), 28, pp. 14–6.
10. An early and important contributor to this endeavour is George S. Rousseau, whose essays have recently been republished in *Nervous Acts: Essays on Literature, Culture and Sensibility* (Basingstoke: Palgrave Macmillan, 2004).
11. See for example Marijke Gijswijt-Hofstra and Roy Porter (eds), *Cultures of Neurasthenia from Beard to the First World War*, The Wellcome Institute Series in the History of Medicine, Clio Medica, 63 (Amsterdam; New York: Rodopi, 2001).
12. For example Stanley W. Jackson, *Melancholia and Depression: From Hippocratic Times to Modern Times* (New Haven; London: Yale University Press, 1986).
13. Michael MacDonald, 'The "Fearefull Estate" of Francis Spira: Narrative, Identity and Emotion in Early Modern England', *Journal of British Studies* (1992), 31, pp. 32–62; idem, *Mystical Bedlam: Madness, Anxiety and Healing in Seventeenth-Century England* (Cambridge: Cambridge University Press, 1981); Michael MacDonald and T.R. Murphy, *Sleepless Souls: Suicide in Early Modern England* (Oxford: Clarendon, 1990).
14. Roy Porter, '"The Hunger of Imagination": Approaching Samuel Johnson's Melancholy' in William F. Bynum, Roy Porter and Michael Shepherd (eds),

The Anatomy of Madness: Essays in the History of Psychiatry, Vol I. *People and Ideas* (London: Routledge, 2004), pp. 62–88.

15. On theories of mind see Roger Smith, 'The History of Psychological Categories', *Studies in the History and Philosophy of the Biological and Biomedical Sciences* (2005), 36, pp. 55–94. Thanks to Rhodri Hayward for this reference. For examples of 'modern' emotional disorders, see Peter Leese, *Shell Shock: Traumatic Neurosis and the British Soldiers of the First World War* (Basingstoke: Palgrave Macmillan, 2002); George L. Mosse, 'Shell-Shock as a Social Disease', *Journal of Contemporary History* (2000), 35, pp. 101–8; Mark S. Micale and Paul Lerner (eds), *Traumatic Pasts: History, Psychiatry, and Trauma in the Modern Age, 1870–1930* (Cambridge; New York: Cambridge University Press, 2001); Ralph Harrington, 'On the Tracks of Trauma: Railway Spine Reconsidered', *Social History of Medicine* (2003), 16 pp. 209–23.
16. Recent work by cultural historians includes Gail Kern Paster Katherine Rowe and Mary Floyd-Wilson (eds) *Reading the Early Modern Passions: Essays in the Cultural History of Emotion* (Philadelphia: University of Philadelphia Press, 2004).
17. Lucien Febvre, 'La sensibilité et l'histoire: Comment reconstituer la vie affective d'autrefois?', *Annales d'histoire sociale* (1941), 3, pp. 5–20, produced in English as 'Sensibility and History: How to Reconstitute the Emotional Life of the Past' in Peter Burke (ed.), *A New Kind of History: From the Writings of Febvre* trans. K. Folca (London: Routledge and Kegan Paul, 1973), pp. 12–26; Peter N. Stearns and Carol Z. Stearns, 'Emotionology: Clarifying the History of Emotions and Emotional Standards', *American Historical Review* (1985), 90, pp. 813–36; idem, *Anger: The Struggle for Emotional Control in America's History* (Chicago; London: Chicago University Press, 1986); Peter N. Stearns, *Jealousy: The Evolution of an Emotion in American History* (New York; London: New York University Press, 1989); idem, *Battleground of Desire: The Struggle for Self-Control in Modern America* (New York; London: New York University Press, 1999).
18. Stearns, *Jealousy*; Stearns and Stearns, *Anger*; Joanna Bourke, *Fear: A Cultural History* (London: Virago Press, 2005).
19. Important examples from anthropology include Jean Briggs, *Never in Anger* (Cambridge, MA: Harvard University Press; London: Oxford University Press, 1970); Robert I. Levy, *Tahitians. Mind and Experience in the Society Islands* (University of Chicago Press: Chicago, London, 1973); Richard A. Shweder and R.A. LeVine (eds), *Culture Theory: Essays on Mind, Self and Emotion* (Cambridge: Cambridge University Press, 1984). On the psychology of emotions, see Jochen Musch and Karl Christoph Klauer (eds), *The Psychology of Evaluation: Affective Processes in Cognition and Emotion* (Mahweh, NJ; London: Erlbaum, 2003); Robert Plutchik, *Emotions and Life: Perspectives from Psychology, Biology and Evolution* (Washington, DC: American Psychological Association, 2003). On the sociology of emotion see David D. Franks and E. Doyle McCarthy (eds), *The Sociology of Emotions: Original Essays and Research Papers*, Contemporary Studies in Sociology, 9 (Greenwich, CT; London: JAI, 1989); Jack Barbalet (ed), *Emotions and Sociology*, Sociological Review Monograph Series (Oxford: Blackwell, 2002); Michael Lewis and Jeanette M. Haviland-Jones (eds), *Handbook of Emotions* (New York: Guildford Press, 2000).
20. For a more detailed historiographical discussion, see Fay Bound, 'Emotion in Early Modern England: Performativity and Practice at the Church Courts of York, c. 1660–1760' (DPhil, York, 2000), pp. 5–6.

21. Norbert Elias, *The Civilizing Process: The History of Manners and State Formation and Civilization*, trans. Edmund Jephcott, 2 vols (Oxford: Blackwell, 1978, 1982).
22. The most famous examples of this trend are Lawrence Stone, *The Family, Sex and Marriage in England, 1500–1800* (London: Weidenfeld and Nicolson, 1977). More recent examples include G.J. Barker-Benfield, *The Culture of Sensibility: Sex and Society in Eighteenth-Century Britain* (Chicago; London: University of Chicago Press, 1992) and Anthony J. Fletcher, *Gender, Sex and Subordination in England, 1500–1800* (New Haven; London: Yale University Press, 1995). For a more detailed critique of this approach see Bound, 'Emotion in Early Modern England', pp. 11–2 and Fay Bound, 'An "Angry and Malicious Mind"?' Narratives of Slander at the Church Courts of York, c. 1660–c. 1760', *History Workshop Journal* (2003), 56, pp. 59–77,
23. Barbara H. Rosenwein, 'Worrying about Emotions in History', Review Essay, *American Historical Review* (2002), 107, pp. 821–45; Bound, 'Emotion in Early Modern England', pp. 8–12.
24. Michael Bamberg, *Language, Concepts and Emotions: The Role of Language in the Construction of Emotions*, Visiting scholars series, Massey University, Department of Psychology, 6 (New Zealand: Department of Psychology, 1996); Patrick Colm Hogan, *The Mind and its Stories: Narrative Universals and Human Emotion*, Studies in Emotion and Social Interaction, 2nd series (Cambridge: Cambridge University Press, 2003); S. Tan (ed.), *Emotion and the Structure of Narrative Film: Film as an Emotion Machine* trans. Barbara Fasting (Hillsdale, NJ: L.Erlbaum Associates, 1996); Marie A. Mills, *Narrative Identity and Dementia: A Study of Autobiographical Memories and Emotions* (Aldershot: Ashgate, 1998) Luisa Passerini, 'Women's Personal Narratives: Myths, Experiences and Emotions' in Personal Narratives Group (eds), *Interpreting Women's Lives: Feminist Theory and Personal Narratives* (Bloomington: Indiana University Press, 1989).
25. R. Harré (ed.), *The Social Construction of Emotions* (Oxford: Basil Blackwell, 1986), Catherine Lutz and Geoffrey M. White, 'The Anthropology of Emotions', *Annual Review of Anthropology* (1986), 15, pp. 405–36.
26. For a discussion, see Judith Butler, *Bodies that Matter: On the Discursive Limits of 'Sex'* (New York: Routledge, 1993), pp. 27–29.
27. On the historical problem of bodily experience, see Lyndal Roper, *Oedipus and the Devil: Witchcraft, Sexuality and Religion in Early Modern Europe* (London: Routledge, 1994), p. 21. See also Butler, *Bodies that Matter*, p. 28.
28. Charles E. Rosenberg, 'Introduction. Framing Disease: Illness, Society and History' in Charles E. Rosenberg and Janet Golden (eds), *Framing Disease: Studies in Cultural History* (New Brunswick, NJ: Rutgers University Press, 1992); David Harley, 'Rhetoric and the Social Construction of Illness and Healing', *Social History of Medicine* (1999), 12, pp. 407–35. See also George S. Rousseau (ed.) *Framing and Imagining Disease in Cultural History* (Basingstoke: Palgrave Macmillan, 2003).
29. Stearns and Stearns, 'Emotionology'.
30. Ibid., p. 813.
31. Rosenwein, 'Worrying', p. 842.
32. Bound, 'Emotion in Early Modern England', pp. 1–19.

33. Penelope Gouk and Helen Hills, 'Towards Histories of Emotions' in Penelope Gouk and Helen Hills (eds), *Representing Emotions: New Connections in the Histories of Art, Music and Medicine* (Aldershot: Ashgate, 2005), p. 26.
34. Bound, 'Emotion in Early Modern England', pp. 24–27; Thomas Dixon, *From Passions to Emotions: The Creation of a Secular Psychological Category* (Cambridge: Cambridge University Press, 2003).
35. Bound, 'Emotion in Early Modern England', pp. 32–34.
36. Rosenwein, 'Worrying about Emotions', p. 844.
37. Pierre Bourdieu, *Outline of a Theory of Practice*, trans. R. Nice, Cambridge Studies in Anthropology, 16 (Cambridge: Cambridge University Press, 1989), especially pp. 72–78; Richard Shusterman (ed.), *Bourdieu: A Critical Reader* (Oxford: Blackwell, 2000); Bound, 'Emotion in Early Modern England', p. 35.
38. Deborah Reed-Danahay, *Locating Bourdieu* (Bloomington, Indiana: Indiana University Press, 2005).
39. Reed-Danahay, *Locating Bourdieu*, p. 2.
40. On the relationship between Bourdieu's concept of 'habitus' and the psychological theories of Norbert Elias, see Reed-Danahay, *Locating Bourdieu*, p. 104.
41. Michelle Z. Rosaldo, 'Toward an Anthropology of Self and Feeling', in Shweder and LeVine (eds), *Culture Theory*, p. 143. See Bound, Emotion in Early Modern England', p. 35.
42. On knowledge production and the influence of science and culture in Bourdieu's writing, see David Swartz, *Culture and Power: The Sociology of Pierre Bourdieu* (Chicago; London: University of Chicago Press, 1997), Chapters 6 and 10.
43. On the complex historical relationship between mind, self and brain, see Joseph LeDoux, Jacek Debiec and Henry Moss (eds), *The Self: From Soul to Brain* (New York: New York Academy of Sciences, 2003). As Thomas Dixon has demonstrated, this shift represented the emergence of 'emotion' as a secular philosophical category by the eighteenth century. See Dixon, *From Passions to Emotions*; H.M. Gardiner, R.G. Metcalfe and J.G. Beele-Center, *Feeling and Emotion: A History of Theories* (New York; Cincinnati: American Book Company, 1987); Bound, 'Emotion in Early Modern England', pp. 29–30.
44. Charles Darwin, *The Expression of Emotions in Man and Animals* (London: John Murray, 1872). Much recent scholarship has focused on this work. See Janet Browne, 'Darwin and the Expression of the Emotions' in David Kohn (ed.), *The Darwinian Heritage* (Princeton; Guildford: Princeton University Press in association with Nova Pacifica, 1985); Mark S. George, 'Reanimating the Face: Early Writings by Duchenne and Darwin on the Neurology of Facial Emotion Expression', *Journal of the History of the Neurosciences* (1994), 3, pp. 21–33; Paul Ekman Joe Campos, Richard Davidson and Frans De Waal (eds), *Emotions Inside Out: 130 Years after Darwin's The Expression of the Emotions in Man and Animals*, Annals of the New York Academy of Sciences, 1000 (New York, NY: New York Academy of Sciences, 2003), Dixon, *From Passions to Emotions*, p. 159.
45. On the context of James's work, see Otniel E. Dror, 'The Scientific Image of Emotion: Experience and Technologies of Emotion', *Configurations* (1999), 7, pp. 355–401.

46. See Carl Georg Lange, *The Emotions*, Psychology Classics, 1 (Baltimore: Williams and Wilkins Company, 1922); Walter B. Cannon, 'The James-Lange Theory of Emotions: A Critical Examination and an Alternative Theory', *American Journal of Psychology* (1927), pp. 106–24.
47. See Randolph R. Cornelius, *The Science of Emotion: Research and Tradition in the Psychology of Emotions* (Upper Saddle River, NJ: Prentice-Hall, 1996), especially pp. 18–57.
48. See also Dixon, *From Passions to Emotions*, passim.
49. Wendy D. Churchill, 'The Medical Practice of the Sexed Body: Women, Men and Disease in Britain, circa 1600–1740', *Social History of Medicine* (2005), 18, pp. 3–22.
50. Sanyantani DasGupta, 'Reading Bodies, Writing Bodies: Self-Reflection and Cultural Criticism in a Narrative Medicine Curriculum', *Literature and Medicine* (2003), 22, pp. 241–256.
51. See John L. Coulehan, 'Tenderness and Steadiness: Emotions in Medical Practice', *Literature and Medicine* (1995), 14, pp. 222–36.
52. Reed-Danahay, *Locating Bourdieu*, pp. 12–3.
53. Ian Burkitt, Social Selves: Theories of the Social Formation of Identity (London: Sage, 1991), p. 90.
54. Bound, 'Emotion in Early Modern England', pp. 10–13.

1

Emotions in the Early Modern Medical Tradition[1]

Fay Bound Alberti

One of the problems facing historians of emotion is the question of 'modernity'. In this specific context this means the well-rehearsed historiographical belief that sometime in the late seventeenth century new ways of thinking and feeling emerged, and emotions in their modern, mental realms were invented.[2] Elsewhere I have argued against historical meta-narratives of emotional change which – primarily drawing on the work of Norbert Elias and the 'civilizing process' – characterize the emergence of 'modern' emotional behaviour as a history of affective restraint versus indulgence.[3] The evidence for this shift is derived from perceived – yet largely unexplored – transformations in medical theory and practice. Historians are accustomed to viewing the post-Restoration period as representing a radical break in medico-scientific ideas about the human body and its capabilities. Despite criticism over the shadows thrown by the Enlightenment, the post-Newtonian age is associated with a revolution in human thought, behaviour and action. As Roy Porter put it, the Enlightenment 'secured the triumph of a radical new rendering of the very constitution of Nature' as Aristotelian metaphysics were overthrown by models of nature 'viewed as matter in motion, governed by laws capable of mathematical expression'.[4] The creation of this mathematical universe underlined the processes by which 'human nature' became discoverable, measurable and quantifiable in ways hitherto unknown.[5]

Many redefinitions of physiological processes took place during the following decades. From humouralism to iatromechanism, through electrical impulses and neuropsychology, the material structure and immaterial potentialities of the human mind and body are believed to have impacted upon, and been influenced by, medico-scientific theories of affect. Yet the development of theories of the emotions – particularly in relation to broader scientific and theological beliefs – is an under-explored area of medical history. This lack is most notable when one considers the importance of emotion theory to histories of philosophy.[6] Most medical history research has focused on emotion as a subject of modern laboratory medicine, rather than in its pre-modern manifestations.[7] Moreover, instead of receiving specific investigation, the emotions have all too often featured as a footnote in physiological investigations concerned with life processes: respiration, digestion and reproduction – and (more recently) the nature of the mind/body relation.[8]

Using a range of published writings by such influential scholars as the Catholic priest and religious controversialist Thomas Wright (c. 1561–1623); the French philosopher René Descartes (1596–1650); the Scottish physician Archibald Pitcairne (1652–1713); the Dutch physician and professor of medicine Hermann Boerhaave (1668–1738); and the Swiss biologist and pioneer physiologist Albrecht von Haller (1708–1777), this chapter explores some of the changing concepts of affect that were characteristic of British and European medico-scientific theory between the seventeenth and eighteenth centuries.[9] It locates individual texts within their historical context, and explores the significance of emotion theories to the complex negotiations of mind and body disputes in which they were embroiled. In so doing it historicizes the transition from mind/body holism towards the splitting of psyche and soma in medico-scientific theory (emotions subsequently being related to the mind in the disciplines of psychology and psychiatry) that has been naturalized in much present day emotion theory.[10] This process was neither linear nor straightforward, the principle of mind/body holism being retained in medical practice well into the modern period. Yet the process by which emotions were increasingly seen to originate in the brain and central nervous system provides an important context for the nineteenth-century drive towards laboratory standardization that has been identified by several contributors to this volume.

Heart and soul: The classical inheritance of humouralism

Until at least the late seventeenth century, interpretations of human emotion focused on the humoural pathology of classic inheritance, the most eloquent or long-lasting exponent of which was Galen of Pergamum (AD c. 129–c. 200). In the 2000 years after Hippocrates, 'Galenism' became a set of connected principles, doctrines and concepts that dominated psychological and physiological theories – including emotion beliefs – throughout much of Europe until at least the eighteenth century.[11] Under Galenism the human body was a 'little world' or microcosm of the universe – the tripartite divisions of heaven, sky and earth corresponding to the three main parts of the human body: the head (reason), the breast (heart) and the lower body (nourishment, procreation).[12] The *soma* therefore possessed all the qualities that made up the 'greater world' of fire, air, water and earth.[13] Four qualities – hot, cold, moist and dry – inhered within these elements: fire was hot and dry; air was hot and moist; water was cold and moist; and earth was hot and dry. In each individual, these characteristics received the form of 'humours', which were believed to course through the body: blood, which was hot and moist like air; choler (or yellow bile) which was hot and dry like fire; phlegm, which was cold and moist like water; and melancholy (or black bile), which was cold and dry like earth.[14]

The proportional balance of these humours within each individual was partly engendered and partly innate, a product of heredity, age, sex, and what contemporaries called the six 'non-naturals': air; food and drink; exercise and rest; sleep and waking; evacuation and repletion; and passions of the soul. Although the passions acted on the spirits and humours, they were also influenced by, and a product of, humoural balance. And an individual's humoural balance was environmental as well as constitutional. As the seventeenth-century English philosopher Thomas Hobbes put it, their proportions 'proceedeth partly from the different constitution of the body, and partly from different Education'.[15] Nevertheless, the overall composition of the humours determined an individual's psychological and emotional demeanour, as humours were produced in the liver and coursed through the veins, mingling with the blood and affecting the mind, soul and body alike as it did so. A disproportionate amount of any of the humours led to constitutional imbalance, including illness, as well as extreme emotional types, and humours and emotions were inseparably linked. In the words of Wright, 'passions engender humors and humors breed passions'.[16]

Although the languages of 'personality types' or 'characters' were not commonly discussed until much later, early modern men and women were understood to display certain psychological tendencies, as expressed in language which we recognize today. A high level of yellow bile made men and women subject to anger (choler) and black bile to sadness (melancholy), while an excess of blood or phlegm made one sanguine (and prone to love-sickness) or phlegmatic. Each age and sex was accorded specific 'prevailing humours' as 'the manners of the soul follow the temperature of the body'.[17] As L. J. Rather put it, 'young men are hot, incontinent and bold, old men are cold covetous and cautious, women are envious, proud and inconstant – and these differences rest on differences in corporeal makeup'.[18]

As this evidence suggests, humouralism was, despite its emphasis on environmental circumstance, inherently gendered. Emotions and passions were often mixed and inter-related. The English divine and scholar Robert Burton (1577–1640) wrote that melancholy could be caused by such diverse passions as sorrow, fear, envy, shame, hatred, anger, ambition and self-love.[19] Yet there were specific emotional tendencies characteristic of each sex. Moreover, those tendencies were apparent in physical differences. Women tended towards a phlegmatic or cold and moist disposition, since their bodies were fleshier, softer and weaker than those of men; their hair longer, their faces paler, and their skin more moist. The greater passivity of women (and later their wombs) also made them more subject to such emotional extremes as hysteria.[20] Men, by contrast, with their leaner bodies and drier complexions, tended to display qualities of courage and anger. This model was also generational, with men in particular being prone to varying emotional behaviours during the course of their lives. In the words of Wright, 'younge men generally are arrogant, prowde, prodigall, incontinent', and old men 'subject to sadnesse caused by their coldness of blood'.[21] Moreover, not only were emotional capacities gendered, but also emotional expressions, such as weeping. The preponderance of water in women's physical constitution meant that women were more prone to tears, and also to sudden, irrational rages, since women's flesh 'is loose, soft and tender, so that the choler being kindled, presently speeds all the body over, and causeth a sudden boyling of the blood about the heart'.[22] Women's anger soon passed, however, since (like old men), they lacked the heat to sustain the emotion.

In these discussions of humoural physiology, emotions were grounded or concretized in the physical body and – more specifically – in the organ of the heart. This is crucial to any understanding of early modern emotion physiology, for while the brain was the seat of reason, the heart was the site of emotion or passion. As Burton put it, this organ was 'the seat and foundation of life, of heat, of spirits, of pulse and respiration, the sun of our body, the king and sole commander of it, the seat and organ of all passions and affections'.[23] In the heat-based economy of emotion physiology, the heart was the centre point where blood was heated, or cooled, respectively, under such passions as anger or fear. Compare Pierre de la Primaudaye's conduct-book description of rage, for instance, with the preaching of the English divine John Downame:

> For first of all when the heart is offended, the bloud boyleth round about it, and the heart is puffed up: whereupon followeth a continuall panting and trembling of the heart and breast. [24]

> [Anger is] an affection, whereby the bloud about the heart being heated, by the apprehension of some injury offered to a man's self or his friends, and that in turn, or in his opinion onely, the appetite is stirred up to take revenge.[25]

This emphasis on the heart as the agent of heating and concoction was compatible with an emotion physiology that linked body, soul and mind in a complex union. It also derived from a more ancient set of emotion beliefs. Eric Jager has shown how Aristotle (heavily influential after the twelfth century in the West) also associated the heart with the vital functions, emotions and sensation, while 'classical Latin commonly used the heart (*cor*) as a synonym for thought, memory, soul and spirit, as well as for the seat of intelligence, volition, character and the emotions'.[26]

Because Renaissance physicians regarded the soul and the body as indivisible, the soul was necessarily involved in the production of emotions as bodily experiences. As 'operations of the soule, bordering upon reason and sense, prosecuting some good thing, or flying some ill thing [and] causing there withall some alteration in the body', emotions were essentially cognitive phenomena. As Wright continued, they were 'movements' of the appetitive faculty (in the mind/soul) that instigated a 'passion' or a change in the humours.[27] This role accorded to the appetitive faculty gave emotions a moral dimension, as to fear or loathe an object was God-given. And it made the physiological process of the passions and their mediating point between *psyche* and *soma* quite logical. The communication of an image via the senses to the brain preceded any judgement about its value; a corporeal alteration in the brain caused 'spirits' to move to the heart, where they would 'signify' the object, and the heart would bend itself to seek or avoid the same.[28] The soul thus had 'power to excite Corporeal Passions *directly*, that is, without considering successively various things', as 'is manifest from her [the soul's] sovereignty over the body, which in all voluntary actions is absolute and uncontrollable'.[29] It is in the context of corporeal alteration that the heart was most evident, for in order to implement the desires of the soul the heart required humours; melancholy blood to produce pain and sadness; blood and choler for anger. The humours were concocted in the heart and the spirits produced were sent around the body to effect the soul's desires. The agitation of the spirits in joy and the free flow of the blood throughout the body were in direct contrast to the physiological experience of fear, when the blood retreated and the soul shrank back from the perceived threat. This process explained the physical manifestations of emotion. As Downame's *Treatise of Anger* continued:

> [Anger] maketh the haire to stand on end, shewing the obdurate inflexiblenesse of the minde. The eyes to stare and candle, as though with the Cockatrice they would kill with their lookes. The teeth to gnash like a furious Bore. The face now red, and soon after pale, as if either it blushed for shame of the mind's follie, or envied others good. The tongue to stammer, as being not able to expresse the rage of the hart. The bloud ready to burst out of the vaines, as though it were affraide to stay in so furious a body. The brest to swell, as being not large enough to containe their anger, and therefore seeketh to ease it selfe, by sending out hot-breathing sighes. The hands to beate the tables and walles, which never offended them. The joyntes to tremble and shake, as if they were afraid of the mines furie. The feete to stamp the guiltlesse earth, as though there were not room enough for it in the whole element of the aire, and therefore sought entrance into the earth also. So that anger deformeth the body from the hayre of the head to the soale of the foote.[30]

If anger caused the blood to boil around the heart, the reverse physiological process was associated with fear, as with grief and sorrow. These 'negative' emotions caused the soul to contract. Thus in grief and sorrow, 'the *Animal Spirits*' were

> recalled inward, but slowly and without violence: so that the *blood* being by degrees destitute of a sufficient influx of them, is transmitted with too slow a motion. Whence the pulse is rendered *little, slow, rare* and *weak*, and there is felt about the heart a certain oppressive *strictness* as if the orifices of it were drawn together, with a manifest *chilness* congealing the blood and communicating itself to the rest of the body.[31]

Such 'dejecting symptoms' had a long–term detrimental effect on the body:

> [extreme sadness] darkneth the *Spirits*, and so dulles the *wit*, obscures the *judgement*, blunts the *memory*, and in a word beclouds the *Lucid* part of the Soul: it doth moreover incrassate the blood by refrigeration, and by that reason immoderately constringe the heart, cause the lamp of life to burn weakly and dimly, corrupt the *nutritive* juice and convert it into that Devil of a Humour, Melancholy. [32]

Then as now, different types of persons experienced greater or lesser degrees of emotion, though the propensity to experience those was explained in physical or spiritual rather than psychological terms. For not only the physical constitution of the body, its humoural composition, but also the soul's misapprehension or misjudgement could be involved, resulting in extreme cases in mania, melancholia or madness. In this context, emotional imbalances – like melancholia – were both mental and physical diseases.

Mechanical philosophy and hydrodynamic principles

The 'new physiology' of the seventeenth century – usually said to have begun with William Harvey's discovery of blood circulation in 1628 – brought considerable changes to medico-scientific theorizing of emotions, as to all aspects of human physiology and psychology.[33] It was not unconnected that new ways of thinking about the human frame – which separated out mind and body more than had been previously conceptualized, and which prioritized the former over the latter as the origin of knowledge and consciousness – coincided with a potential secularization of the body and the emergence of rational philosophy as a way of understanding how humans thought and felt. The influence of Isaac Newton in this endeavour is well known. Newton's *Philosophiae naturalis principia mathematica* (*Mathematical Principles of Natural Philosophy*), published in 1687, laid out a new science of dynamics that would be applied to many other fields, including science and medicine.[34]

The appeal of Newton's method lay in providing a mathematical approach that could be applied to rational mechanics. It gave rise to various schools of rational medicine, including iatromathematics, iatromechanics and iatrochemistry, as well as contributing to Cartesian principles and mechanical philosophy.[35] Indeed, the rationalist physician

George Cheyne (1671/2–1743), author of such influential works as *The English Malady*, argued in 1701 that what medicine needed above all else was a '*Principia medicinae theologia mathematica* based on Newton's *Principia* – one that would integrate medicine and mathematics' beyond all doubt.[36] Debates between rationalists (who developed systems of thought based on first principles and theories), and empiricists (who derived their findings and theories from observation and analysis) continued throughout the eighteenth century.[37]

While theologians and moral philosophers discussed the associative and cognitive dimensions of emotions, understandings of their function and structure were influenced by medico-scientific debates on the material structure of the body. This included the iatrochemical approach of Thomas Willis (1621–1675); the mechanical physiology of Pitcairne, well-known as a physician and mathematician throughout Europe, and of Friedrich Hoffman (1660–1742); the hydro-dynamical circulatory physiology of Boerhaave; the models of sensibility and irritability found in the work of von Haller; and the neurophysiology of William Cullen (1710–1790). The combined impact of their research helped to redefine relations between corporeal and incorporeal functions at the level of theory, and facilitated a concept of the body that operated according to distinct laws of motion, with or without the presence of a soul. Feeling and movement (sensibility and irritability) were instead postulated as the phenomena characterizing life.[38] These shifts necessarily influenced broader meta-physical considerations about the nature of emotion and of the mind/soul relation.

Mechanistic physiology was pioneered in France by the French philosopher and experimentalist Pierre Gassendi (1592–1655); the mathematician and theologian Marin Mersenne (1588–1648); and especially by René Descartes (1596–1650) whose *Les passions de l'âme* offered mechanistic interpretations of human sensation according to laws of matter and motion.[39] Descartes' work is often believed to have laid the foundations for modern neurophysiological psychology, and he certainly set out the mind as a separate focus for study in ways that were hitherto unrealized. Although Descartes reaffirmed traditional physiological explanations of the passions by retaining the animal spirits of the nervous system as a key factor, he introduced the pineal gland as the material site for the interaction of the soul with the body, and gave mechanical explanations in place of traditional humoural notions.[40] In focusing on the doctrine of the reflex – which provided a rationale for the machine-like functioning of the nervous system and the body it controlled – Descartes applied mechanical philosophy to the functioning of the human body in ways that potentially removed the soul from its position as a causal factor in bodily changes, and made it no more than a passive observer.[41] In this model, the soul was independent of, and separate from, the corporeal body, though both were mutually influential through a process of interaction.

By separating the 'principle of life', common to all animals, from the 'principle of mind', by placing limitations on the mind's influence on the body and by presenting a distinction between mental and bodily processes in a way that potentially divorced physiological and psychological principles, Descartes offered a physical interpretation of the origin of passions which was distinct from the approach of Thomas Wright, discussed above.[42] Consider the following synopsis of Descartes' position, as given by Susan James in her erudite work on *Passion and Action.*[43] Firstly, the animal spirits would move along the nerves from the sense-organs to the brain. In the brain, motions in the cerebral cavities would push the spirits along the nerves, causing such bodily events as a rush of blood to the area around the heart, or the contraction of the muscles in the limbs. This instinctive physical motion explains the behaviour of animals (such as sheep that run from a wolf without feeling fear), for which passions are unsophisticated reflex mechanisms because they lack souls. In humans, by contrast, the motions of the cerebral cavity may move the pineal gland (identified by Descartes as the site of interaction between mind/soul and body), and thereby cause perceptions in the soul that may be sensory representations of passions. The same process was responsible for the physical effects associated with specific emotions, such as blushing in shame or paling in fear. Instead of being impulses of the mind/soul that were subsequently felt in the body, Descartes presented emotions as bodily happenings felt first in the *soma* and then presented to the *psyche* through the senses and processes of feeling and understanding located in the brain. In these terms they were 'passions of the mind' – as perceptions perceived by the mind and related to the mind – but, as Rather explains, 'caused, mediated and strengthened by movements of the nervous fluid'.[44]

Descartes' articulation of emotions as bodily experiences did not end debates over the role of the soul.[45] Nor did it mean that the soul did not continue to be understood in many quarters as the originator of the passions and the mediator between *psyche* and *soma*. Following Boerhaave, some mechanists held the soul to be a rational, immaterial principle which was somehow attached to the material body, and which operated in accordance with established rules on matter and motion.[46] Vitalists and animists alike rejected the limited status this accorded to the soul, claiming that rather than being an adjunct of the body (its physiological role restricted to 'willed activity and to consciously perceived sensations'), the soul or *anima* acted independently.[47] Indeed, the soul performed the ordinary functions of life in humans, although those of lower animals might be performed by mechanical principles. For the animist Georg Ernst Stahl (1659–1734), court physician to Frederick William I and professor at the University of Halle, therefore, 'the soul' (a term he used to cover *all* perceptual processes: sensory, mental and emotional), formed images and ideas, incited affective responses, and effected physiological change within the body. Stahl's study of the physiological impact of emotional responses led him to assert that the body followed the soul, an idea that could not be *disproved* medically, as the actual relation between emotional experience and physical expression remained uncertain. Moreover, in rejecting the idea of the body as a mechanical automaton, Stahl regarded it as an 'inherently unstable instrument of life, constantly in need of regulation and repair by the soul'.[48]

As Geyer-Kordesch has observed, this unity of body and soul 'directly challenged a somatically oriented medicine as well as post-Cartesian philosophy'.[49] There was therefore no separation of mind and body, or reason and emotion in the work of Stahl: emotion was connected to reason and imagination; the pre-Cartesian model of identity remained intact. Later in the century, vitalism offered a more sophisticated approach, largely by replacing an abstract soul with the doctrine of sensibility, as seen in the works of the French physicians Paul Joseph Barthez (1734–1806) and Théophile Bordeu (1722–1776). While Barthez wrote of a 'vital principle' which differed from the thinking mind and was the cause of life itself (a principle moreover which differed from the soul and was found not only in humans but also in animals and plants), Bordeu's contribution was to attribute glandular activity and other vital functions to 'sensibility', a vital force that resided in the material body and was not imposed on it from without.[50]

If Cartesian philosophy did not end debates over the role of the soul in emotional experience, it did provide the basis for alternative interpretations of human physiology to those provided by humouralism. As such, it provided a shift in physiological interpretation that was compatible with parallel developments in philosophy and physiology, including William Harvey's discovery of the circulation of the blood, the association principles of John Locke and others, and the development of nerve theory.[51]

Descartes' work on emotions therefore needs to be understood in the context of medico-scientific change generally. In providing a way to think about emotions less in terms of humoural balance and more in accordance with the clockwork functioning of the body, mechanical philosophy popularized a language of ebb and flow in which new and modified versions of human physiology could flourish.

The late seventeenth century saw the emergence of hydro-dynamical theories of the body and mind. Scottish physicians like Pitcairne and George Cheyne, Pitcairne's pupil who was equally well known for his attempts to practise what he preached in the field of medicine, conceived of the human body as a system of 'pulleys, springs and levers, pipes and vessels, its fluids being governed by the laws of hydraulics'.[52] For Pitcairne, the living body was composed of 'canals' and conveyed 'fluids' which moved with blood circulation: bodily disease could be viewed as deriving from an unusual circulation of blood in some or all of the parts of the body.[53] In this context it was possible to perceive emotional extremes, such as melancholia, as the product of defective motions of the blood and its excessive accumulation in the brain, largely as a result of the affected flow and distribution of the animal spirits.[54] Again, the material construction of the body, rather than any psychological determinants, could be the originator of emotional experience.

Demonstrating the common practice of uniting older and newer theories of human physiology to formulate revised interpretative theories, Friedrich Hoffman used a notion of fermentation (characteristic of iatrochemical discussions) in conjunction with a reliance on mechanical models. In his work the human body was conceived as a machine with solid and fluid parts. Although obstructions in the solids could interfere with the equilibrium of motion, it was the movement of the fluids and their constitution that determined health, disease and emotional tendencies: 'in melancholics the spirits are indistinct and fixed, and approach a sort of acid nature. They not only leave enduring fixed ideas in the brain pores, but promptly uncover similar traces, of ideas of sadness, terror, fear, and so on'. When the body fluid became acidic and fixed, therefore, an individual subsequently became 'slow, timid, and sad'.[55]

As Jackson has argued, Hoffman's influence in physiological theory increased over the several decades from the 1690s, an influence matched only by Hermann Boerhaave (whose work Hoffman did much to systematize) in the eighteenth century. Boerhaave developed a system by which most eighteenth-century medico-scientific research into emotions was conducted.[56] Indeed, the Edinburgh medical school was established in 1726 along the lines of the system in place at Leyden, where Boerhaave was influential.[57] Boerhaave viewed forces as inherent in matter, expressed as mechanical movements and determined by mass, number and weight. Forces were similarly expressed within the body by movement, contraction and relaxation. According to hydrodynamical principles of motion, Boerhaave constructed a system of physiology that viewed the body as a machine containing solid parts as its framework and circulating fluids that were charged by mechanical forces.[58] Health and disease were dependent on the circulation of the fluids around the body, as were extremes of emotional states. Hypochondriacal disease or melancholia, for instance, derived from the thickening of the fluids, and the accumulation of pathogenic material in the body from the bowels of the abdomen to the chest. As the vessels became obstructed and the brain received putrefied matter, 'the Vessels of the abdominal Bowels create a Stagnation, Alteration and Accumulation of black Choler which insensibly increaseth, though the Body was very healthful but a little before: And also that the same black Choler, when bred from bodily Causes, doth produce that Delirium'.[59]

The role of the nerves

The relevance here of pathological theories of emotion as linked to the material structure of the human body is that mechanistic and iatrochemical and hydro– dynamical theories of the body moved away from humoural interpretations but continued to explain emotional experience in materialistic terms. Although the circulation of fluids round the body still held important meanings as under humouralism, eighteenth–century theorists reconceptualized the body in which the fluids flowed. As the human frame became a composite of fluids and solids, it was the solids (nerves and fibres) that were conceived as the 'true basis of the body'.[60] Rather than being related to the apprehension of the soul *as* mind, emotions were redefined as a product of sensory perception and material processes. Nervous physiology allowed the main office of the brain and nerves to be 'sensation; that is, to suffer changes from the impressions of external substances in the parts of the body affected by them, and to undergo analogous changes in the representations of the mind'.[61]

This is not the place to revisit the extensive literature on nerve theory and the culture of sensibility in the eighteenth century.[62] What is relevant here is that theories of the nerves began to explain emotional susceptibilities and disorders. As William Clark, MD (1697/8–*c.* 1780),

Intra-Licentiate of the College of Physicians in London put it in 1727, 'every Faculty of the Mind depends on the nervous system', and 'all the Functions of the Body, rightly exerted, depend upon a due distribution of the nervous Fluid'.[63] In this context, the physiology of fear, for instance, was attributable to an increase in nervous fluid, which contracted the muscles and signalled alterations in the motion of the blood and the heart.[64] The exertion of the muscles in anger, or the 'palpitation of the heart in terror' similarly derived from 'exertion of the nervous power', the strength of which depended on the state of the nerves. Since the quality of the solids and the circulation of the fluids dictated mental and bodily health, obstructions in either produced illness because of interdependence between connecting parts. For this reason the physician and natural philosopher Robert Whytt (1714–1766) famously claimed that 'all diseases may, in some sense, be called affections of the nervous system, because in almost every disease the nerves are more or less hurt; and, in consequence of this, various sensations, motions and changes, are produced in the body'.[65]

These developments in the field of medico-scientific theory were linked to developments in moral philosophy and early psychology that made it possible for emotions to be viewed as primarily physical phenomena. The external senses – sight, smell, hearing, taste, touch – transmitted ideas of 'external substances' to the mind, the 'nervous power' instigated to respond being responsible for the effects of the passions:

> which, if lively and exhilarating, relax the influx of blood; and by the remission of the nerves; hence redness, moisture and turgescence of the skin. Those passions, which are languid and depressing, contract the exhaling vessels; as appears from the dryness of the skin, produced by them: from the goose-skin, by terror; and from diarrhoea, caused by fear.[66]

The ability of emotions to cause physical illness was explicable because 'they also seem to dilate the inhaling vessels, whence fear facilitates the action of the smallpox and the plague'.[67] Whereas the external senses 'being affected by external objects', transmit 'some change by the nervous spirits' to the brain, the internal senses – including thought, imagination and memories – were similarly responsive to external qualities (being 'impressed in the body itself, and indeed in the medulla of the brain') by objects of sight, smell, hearing, taste or touch.[68] A lack of judgement or irrational passion could be provoked by the lack of a 'healthy constitution of the brain.

For when that is compressed, irritated, exhausted of blood, or changed in its fabric', the faculty of judgement failed.[69]

It would be naïve to suggest any wholesale transition of emotion theory from mental to corporeal causation during the long eighteenth century, or back again during the nineteenth. In classical medicine it is possible to find physicians ascribing emotional states to physical causes – as in Hippocratic assertions that the heart might become filled with blood because of impeded menstruation, thus causing stagnation and putrefaction of the blood, suicidal thoughts, fear and distress.[70] Moreover, under nervous physiology there remained the possibility of some passions being linked to mental apprehensions, some essence of understanding that was distinct from material processes. Von Haller therefore allowed that there were 'affections of the human mind' that were linked to some innate perception of what was beneficial or detrimental to an individual's safety or health: 'the presence of good constitutes joy; the desire of good, love; the expectation of good, hope; the presence of evil, sorrow, terror or despair; the dislike of evil, hatred; and the expectation of evil, fear', all of which were somehow 'unconnected with the properties of matter, or certainly less simple, understood or mechanical' than other passions.[71] Other medics, including Dr Corp, MD of Bath, abstractedly, and without explanation, wrote of certain emotions – 'Hope, Joy, Anger, Fear, Grief, Anxiety' – as '*purely mental*, as originating in the Mind, and not excited by, or blended with, corporeal sensation.'[72]

Although there continued to be a school of thought in eighteenth-century medical culture that viewed emotion as primarily a mental phenomenon with bodily effects (the 'common sense' model with which we are most familiar), most viewed emotions as material entities, produced by the condition and structure of the *soma*. Their emphasis on the body's constitution meant that an individual's susceptibility to degrees of emotion differed according to their physical predisposition, mental apprehension being dependent on the physical sensations. To this end the influential physician Thomas Cogan (1736–1818), published a five-volume *Treatise on the Passions and Affections of the Mind* (1813) which asserted that passions derived from some 'originating cause' (perception or idea) that 'violently agitates the corporeal frame' and causes a 'change in the state and disposition of the mind'.[73] The belief that emotional experience was linked to corporeal characteristics led to a view of emotional types that were identifiable not in terms of an individual's humoural propensity, but according to the state of his or her nerves and fibres. Those with a 'large brain and thick strong nerves' possessed 'a great sensibility as well of the whole body as of the organs of sense. Hence arises a ready apprehension of objects, and an increase of understanding and knowledge ... which choleric persons possess in so eminent a degree; but along with this condition of the nerves, they are excessively liable to grief and anger'. By contrast, according to von Haller, those with a 'small brain and slender nerves' possessed 'senses more dull and a phlegmatic–melancholic torpor conjoined'.[74]

Emotional pre-disposition could be impacted by environmental factors that modified the condition of the nervous system. Much has been written about the class-based diagnoses of hypochondria as an affliction, and the influence of habits of living – emphasizing the 'non-naturals' – was stressed in literature on emotion. Thus 'persons of gross full habits, the robust, the luxurious, the drunken, and they who sup late' were most likely to succumb to excess fears and even nightmares.[75] Of course, the gendering of hypochondria was also significant. Like those of children, the fibres of women were generally 'lax and soft, the nerves extremely irritable, and the fluids thin' [76] While old age brought a rigidity of the fibres for both sexes and an insensibility of the nerves – giving rise to very different diseases (and treatments) to those that attacked the young, – female constitutions, ordinarily prone to as 'peculiar sensibility of the nervous system', were regularly disturbed by menstruation.[77] Nervous debility and emotional complaints frequently plagued 'women who are obstructed; Girls of full, lax habits, before the eruption of the Menses'.[78] This was especially the case during menopause, for 'when Women pass the fruitful seasons of life, and the delicate uterine Tubes contracting themselves, become too rigid, and resist the impetus of the Fluids so as to prevent the usual discharges'.[79]

* * *

This chapter has demonstrated that there was nothing straightforward about medical theories of emotion in the seventeenth and eighteenth centuries, or about the prioritising of the brain in modern emotion theories.[80] What we find repeatedly in the medical history of emotion is rather the opening up of new spaces by which emotion physiology and psychology was theorized and conceptualized. Theories of emotion did not take place in a vacuum, or as a result of some unmediated scientific 'progress'. Rather, they were consistently influenced by broader debates about the respective roles of God, humans and animals, about the relationship between mind, body and soul and about the social function of ethics and morality. For reasons of space I have not dealt with the language of emotion and the passions, or with theological perspectives on emotion, though these issues have been considered elsewhere.[81] One of the most important developments taking place as a consequence of the 'new physiology', I have argued, was the emergence of the nerves and fibres – over the heart and fluids – as the site of emotion activity. Part of this development was that more emphasis was placed on the material structure of the body in the eighteenth century than in the seventeenth as the cause of emotional extremes. By the mid-nineteenth century, moreover, as Stephen Jacyna has shown, debates on the material causation of mental and emotional instability would also influence developments in neurological investigations and theories of mind.[82] That emotion could be viewed as part of the mechanistic structure of the human body, rather than as a product of the soul, meant that it was possible to produce alternative, secularized interpretations of affect to those that previously existed.

This did not, however, spell the end for humoural pathology, or for traditional accounts of mind/body holism. In the main, new and emerging theories were accommodated into, rather than replacing, traditional pre–existing theories of the *psyche* and the *soma*.[83] In most medical treatises of the period, the nerves were viewed less as 'strings' that transported sensation through the body, moreover, than as conduits for the animal spirits or humours. Thus in Henry Nicholson's *A Brief Treatise of the Anatomy of Humane Bodies* (1709), nerves are described as 'a complication of the finest and most subtile fibers, fill'd with the most volatile Spirit, to receive matter' as it circulated around the body. This concept is clearly closely related to that of the humours as the determining fluid that coursed around the body.[84] Not until the nineteenth century was the alliance between medicine, Lockean philosophy of the mind and the 'science of man' fully realized.[85] The social and institutional processes by which that realization took place are among the themes raised in the following essays.

Notes

1. The research for this chapter was made possible by the generous funding of the Wellcome Trust.
2. For a recent discussion see Thomas Dixon, *From Passions to Emotions: The Creation of a Secular Psychological Category* (Cambridge: Cambridge University Press, 2003), pp. 20–1.
3. Norbert Elias, *The Civilizing Process: The History of Manners and State Formation and Civilization*, trans. Edmund Jephcott, 2 vols (Oxford: Blackwell, 1978, 1982). See Fay Bound, 'Emotion in Early Modern England, 1660–1760: Performativity and Practice at the Church Courts of York' (DPhil, University of York, 2000); eadem, 'Writing the Self? Love and the Letter in England c. 1660–c. 1760', *Literature and History* (2002), 11, pp. 1–19; eadem, 'An "Angry

and Malicious Mind"? Narratives of Slander at the Church Courts of York, c. 1660–c. 1760', *History Workshop Journal* (2003), 56, pp. 59–77. See also Barbara H. Rosenwein, 'Worrying about Emotions in History' Review Essay, *The American Historical Review* (2002), 107, pp. 821–45.

4. Roy Porter, *Enlightenment: Britain and the Creation of the Modern World* (London: Penguin, 2000), p. 138.
5. See William Clark, Jan Golinski and Simon Schaffer (eds.), *The Sciences in Enlightened Europe* (Chicago; London: University of Chicago Press, 1999), Part 3; Richard Olsen, *The Emergence of the Social Sciences 1642–1792* (New York: Twayne Publishers; Oxford: Macmillan, 1993), Chapters 3 and 4; Christopher Fox, Roy Porter and Robert Wokler (eds), *Inventing Human Science: Eighteenth-Century Domains* (Berkeley: University of California Press, 1995).
6. See Dixon, *From Passions to Emotions*, p. 5. A recent important contribution to early modern medical philosophy is Harold J. Cook, 'Body and Passions: Materialism and the Early Modern State', *Osiris* (2002), 17, pp. 25–48. Historiographical examples of the philosophy of the passions in the early modern period include Katherine Park, 'The Organic Soul' in Charles B. Schmitt, Quentin Skinner, Eckhard Kessler and Jill Kraye (eds), *The Cambridge History of Renaissance Philosophy* (Cambridge: Cambridge University Press, 1988), pp. 464–84; Anthony Levi, *French Moralists: The Theory of the Passions 1585–1649* (Oxford: Clarendon Press, 1964); Susan James, *Passion and Action: The Emotions in Seventeenth-Century Philosophy* (Oxford: Clarendon Press, 1997).
7. Otniel E. Dror, 'The Scientific Image of Emotion: Experience and Technologies of Inscription', *Configurations* (1999), 7, pp. 355–401; idem, 'Counting the Affects: Discoursing in Numbers', *Social Research* (2001), 68, pp. 357–78.
8. Examples include Andrew Gregory, *Harvey's Heart: The Discovery of Blood Circulation* (Cambridge: Icon Books, 2001); Robert G. Frank, *Harvey and the Oxford Physiologists: A Study of Scientific Ideas and Social Interaction* (Los Angeles: University of California, 1978); Gerald L. Geison, *Michael Foster and the Cambridge School of Physiology: The Scientific Enterprise in Late Victorian Society* (Princeton, NJ: Princeton University Press, 1978); Stanley Finger, *Minds Behind the Brain: A History of the Pioneers and their Discoveries* (Oxford: Oxford University Press, 2000).
9. While modern definitions of the 'theological' or the 'scientific' do not strictly pertain here, this article's focus on physiological process uses a rather different source base, and an earlier chronology, than that used by Thomas Dixon in his focus on philosophical and psychological categories. See *From Passions to Emotions*, Introduction.
10. For reasons of the relatively recent origin of psychology and psychiatry as disciplines, their historiographies are not considered in this chapter. For an introduction see George S. Rousseau, 'Psychology' in idem (ed.), *The Ferment of Knowledge: Studies in the Historiography of Eighteenth-Century Science* (Cambridge: Cambridge University Press, 1980), Dixon, *From Passions to Emotions*, introduction.
11. See Galen, *On the Passions and Errors of the Soul,* trans. P.W. Harkins ([Columbus]: Ohio State University Press, 1963): O. Temkin, *Galenism: Rise and Decline of a Medical Philosophy* (Ithaca, NY; London: Cornell University Press, 1973), Chapter 1.

12. See Helkiah Crooke, MD, *Mikrokosmographia: A Description of the Body of Man* (London: Printed for William Laggard, 1615) and the discussion in J.B. Bamborough, *The Little World of Man* (London; New York: Longmans, Green, 1952), Chapter 1.
13. For a good, if dated, synopsis see Lily Campbell, *Shakespeare's Tragic Heroes: Slaves of Passion* (London: Methuen, 1930), p. 51.
14. Temkin, *Galenism*, Chapter 1. For one interpretation of this fluid economy in action, see Ulinka Rublack, 'Fluxes: The Early Modern Body and the Emotions', trans. Pamela Selwyn, *History Workshop Journal* (2002), 53, pp. 1–17.
15. Thomas Hobbes, *Leviathan*, ed. R. Tuck (1651, repr. Cambridge: Cambridge University Press, 1991), p. 53.
16. Thomas Wright, *Passions of the Minde in Generall* (1601, 1604, repr. Urbana: University of Illinois Press, 1971), p. 64.
17. Wright, Passions of the Minde, p. 40.
18. L.J. Rather, 'Old and New Views of the Emotions and Bodily Changes: Wright and Harvey versus Descartes, James and Cannon', *Clio Medica* (1965), 1, pp. 1–25, 4.
19. Robert Burton, *Anatomy of Melancholy* (1621, repr. New York: New York Review of Books, 2001), Part 1. See also Akhito Suzuki, 'Mind and its Disease in Enlightenment British Medicine' (PhD, University College London, 1992) p. 53.
20. Bamborough, *Little World of Man*, p. 64.
21. Wright, *Passions of the Minde*, pp. 71–3.
22. Levinus Lemnius, *The Secret Miracles of Nature*, (London: Jo Streater, 1658), p. 274 cited in Bamborough, *Little World of Man*, p. 64.
23. Burton, *Anatomy of Melancholy*, pp. 152–3.
24. Pierre de la Primaudaye, *The French Academie,* trans. Thomas Bowes, Richard Dolman and W.P. (London: Printed for T. Adams, 1618 [1577]), p. 471.
25. John Downame, *A Treatise of Anger* (London: Printed by T. E. for William Welby, 1609), p. 3.
26. Eric Jager, *The Book of the Heart* (Chicago; London: University of Chicago Press, 2000).
27. Wright, *Passions of the Minde*, pp. 11, 8.
28. See the discussion in Rather, 'Old and New Views of the Emotions and Bodily Changes', p. 4.
29. Walter Charleton, *Natural History of the Passions* ([London] In the Savoy: Printed by T. N. for J. Magnes, 1674), p. 70.
30. Downame, *Treatise of Anger*, p. 56.
31. Charleton, *Natural History*, p. 151. On the 'animal spirits' see L.S. Jacyna, 'Animal Spirits and Eighteenth-Century British Medicine' in Yosio Kawakita, Shizu Sakai and Yasuo Otsuka (eds), *The Comparison Between Concepts of Life-Breath in East and West*, (Tokyo; St. Louis: Ishiyaku EuroAmerica, 1995).
32. Charleton, *Natural History*, p. 151.
33. William Harvey, *Exercitatio antomica de motu cordis et sanguinus in animalibus* (Frankfurt, 1628); Gregory, Harvey's Heart.
34. Isaac Newton, *Philosophiae Naturalis Principia Mathematica* (London: J. Streater for Royal Society, 1687); Niccolò Guicciardini, *Reading* the Principia: *The Debate on Newton's Mathematical Methods for Natural Philosophy from 1687 to 1736* (Cambridge: Cambridge University Press, 1999).

35. See George S. Rousseau, *Enlightenment Borders: Pre- and Post-Modern Discourses. Medical and Scientific* (Manchester: Manchester University Press, 1991), p. 125.
36. Rousseau, *Enlightenment Borders*, p. 84.
37. On the difficulties of making clean-cut distinctions between practitioners in each of these fields, see Rousseau, *Enlightenment Borders*, Chapter 5, esp. p. 120.
38. See I.A. Bowman, 'William Cullen (1710–1790) and the Primacy of the Nervous System', (PhD, University of Indiana, 1975), p. 12.
39. René Descartes, *Les passions de l'âme* (Amsterdam [Henry Le Gras] chez Louys Elzevier, 1649). See Richard Olson, *The Emergence of the Social Sciences 1642–1792* (New York: Twayne Publishers; Oxford: Macmillan, 1993), p. 39; Roy Porter, *Enlightenment*, p. 139.
40. See Stanley W. Jackson, *Melancholia and Depression: From Hippocratic Times to Modern Times* (New Haven; London: Yale University Press, 1986), p. 21.
41. See Rather, 'Old and New', p. 1.
42. On the 'goodness of the passions' in Descartes, see Cook, 'Body and Passions', p. 33.
43. James, *Passion and Action*, pp. 96–7.
44. Rather, 'Old and New', p. 12.
45. Cook, 'Body and Passions', p. 35.
46. See Elizabeth L. Haigh, 'Vitalism, the Soul and Sensibility: The Physiology of Théophile Bordeu', *Journal of the History of Medicine and Allied Sciences* (1976), 1, pp. 30–41.
47. Haigh, 'Vitalism', p. 30.
48. John P. Wright and Paul Potter (eds), *Psyche and Soma: Physicians and Metaphysicians on the Mind-Body Problem from Antiquity to Enlightenment* (Oxford: Clarendon Press, 2000), p. 10.
49. See Johanna Geyer-Kordesch, 'Georg Ernst Stahl's Radical Pietist Medicine and its Influence on the German Enlightenment', in Andrew Cunningham and Roger French (eds) *The Medical Enlightenment of the Eighteenth Century* (Cambridge; New York: Cambridge University Press, 1990), p. 68.
50. Haigh, 'Vitalism', p. 31.
51. Harvey, *De motu cordis*; John Locke, *An Essay Concerning Human Understanding*, ed. A.S. Pringle-Pattison; Introduction by Diané Collinson, Wordsworth Classics of World Literature (Ware: Wordworth Editions, 1998 [1697]). Much has been written about the domination of nerve theory in eighteenth-century diagnoses. For reprints of his now classic essays on the subject, see George S. Rousseau, *Nervous Acts: Essays on Literature, Culture and Sensibility* (Basingstoke; New York: Palgrave Macmillan, 2004).
52. Porter, *Enlightenment*, p. 139. See also Roy Porter (ed.), *George Cheyne: The English Malady* (1733, facs. repr. London: Tavistock/Routledge, 1991). On Cheyne's struggle with his weight and attempts to live rationally, see Rousseau, *Enlightenment Borders*, Chapter 4.
53. See Jackson, *Melancholia and Depression*, p. 117.
54. Ibid., p. 288.
55. Ibid., p. 118.
56. For an introduction to Boerhaave's work see Andrew Cunningham, 'Medicine to Calm the Mind: Boerhaave's Medical System, and Why It Was Adopted in Edinburgh' in Cunningham and French (eds) *The Medical Enlightenment*, p. 40.

57. Cunningham, 'Medicine to Calm the Mind', pp. 43, 65.
58. Jackson, *Melancholia and Depression*, p. 119.
59. Quoted in ibid., p. 120.
60. A. Von Haller, *First Lines of Physiology* 2 vols, ed. Lester S. King (1786, facs. repr. New York: Johnson Reprint Corp, 1966), 1, p. 9.
61. Von Haller, *First lines*, vol.1, p. 201.
62. On the pervasiveness of nervousness and sensibility as cultural paradigms, see Rousseau, *Nervous Acts*, introduction; G.J. Barker-Benfield, *The Culture of Sensibility: Sex and Society in Eighteenth-Century Britain* (Chicago; London: University of Chicago Press, 1992); Janet Todd, *Sensibility: An Introduction* (London: Methuen, 1986); Sydney McMillen Conger (ed.), *Sensibility in Transformation: Creative Resistance to Sentiment from the Augustans to the Romantics* (Rutherford: Fairleigh Dickinson University Press; London; Cranbury, NJ: Associated University Presses, 1990).
63. William Clark, MD, *A Medical Dissertation Concerning the Effects of the Passions on Human Bodies* (Bath; London: for W. Frederick, 1752), pp. 38–40.
64. Clark, *Medical Dissertation*, p. 40.
65. Robert Whytt, *Observations on the Nature, Causes and Cure of those Disorders Which Have Been Commonly Called Nervous, Hypochondriac or Hysteric* (London: T. Maiden, 1797), p. 42. On Whytt see R.K. French, *Robert Whytt, the Soul and Medicine* (London: Wellcome Institute of the History of Medicine, 1969).
66. Von Haller, *First Lines*, vol. 1, p. 201.
67. Ibid., p. 214.
68. Ibid., pp. 271–2.
69. Ibid., p. 275.
70. See Beate Gundert, 'Soma and Psyche in Hippocratic Medicine' in Wright and Potter (eds), *Psyche and Soma*, p. 28.
71. Von Haller, *First Lines*, vol. 1, p. 276.
72. Dr Corp, MD of Bath, *An Essay on the Changes Produced in the Body by the Operations of the Mind* (London: James Ridgway, 1791).
73. T. Cogan, MD, *A Treatise on the Passions and Affections of the Mind*, 5 vols (London: T. Cadell and W. Davies, 1813), vol. 2, Part 1: 'On the Pursuit of Well-Being', p. 6.
74. Von Haller, *First Lines*, vol. 1, p. 85.
75. John Bond, MD, *An Essay on the Incubus or Night-Mare* (London: D. Wilson & T. Durham, 1753), p. 5.
76. William Buchan, *Domestic Medicine: or, a Treatise on the Prevention and Cure of Diseases by Regimen and Simple Medicines* (1772 edition, facs. repr. New York; London: Garland, 1985), p. 168.
77. Buchan, *Domestic Medicine*, p. 168, William Rowley, *A Treatise on Female, Nervous, Hysterical, Hypochondriacal, Bilious, Convulsive Diseases; Apoplexy and Palsy, with Thoughts on Madness, Suicide, &c* (London: C. Nourse, E. Newbery and T. Hookham, 1788), p. 1.
78. Bond, *Essay on the Incubus*, p. 46.
79. Ibid., p. 50.
80. For an important work on developments in neurology and theories of brain function in the modern period see L.S. Jacyna, *Lost Words: Narratives of Language and the Brain, 1825–1926* (Princeton, NJ; Oxford: Princeton University Press, 2000).

81. Bound, 'Emotion in Early Modern England', Dixon, *From Passions to Emotions*, *passim*.
82. See L.S. Jacyna, 'Somatic Theories of Mind and the Interests of Medicine in Britain, 1850–1879', *Medical History* (1982), 26, pp. 233–58.
83. On the preservation of traditional theological and philosophical discourses about the soul and the mind 'right up to the modern day', see Dixon, *From Passions to Emotions*, p. 21.
84. Henry Nicholson, MD, *A Brief Treatise of the Anatomy of Humane Bodies* (London, 1709), p. 5. See also J. Purcell, *A Treatise of Vapours* (London: Printed for Edward Place, 1707); Christian Uvedale, *The Construction of the Nerves and the Causes of Nervous Disorders Practically Explained*, 2nd ed. (London: Printed for R. Baldwin, 1758); Richard Blackmore, *A Treatise of the Spleen and Vapours* (London: J. Pemberton, 1725).
85. Roy Porter, *Medicine in the Enlightenment* (Amsterdam: Rodopi, 1995), p. 337.

2

Patients and Passions: Languages of Medicine and Emotion, 1789–1850[1]

Thomas Dixon

The examination of the King's physicians

At the end of November 1788, the formidable 'Dr Willis, a physician of … peculiar skill and practice in intellectual maladies' was summoned to the royal palace at Kew.[2] This invitation, issued reluctantly by Queen Charlotte, was a last resort. It had become increasingly clear that King George was sinking into insanity, and that the royal physicians were either unable or (understandably) unwilling to exercise the requisite medical authority over their patient and sovereign. The King's bouts of delusional raving continued to get worse. So it was that the controversial Dr Francis Willis, the keeper of a mad-house in Lincolnshire, came to meet the King at Kew on 5 December 1788. On the occasion of this first encounter, the King, noticing Dr Willis's clerical attire, asked him whether he was in the employment of the church. Willis replied that he had been formerly but now 'attended chiefly to physick'. The King's equerry Robert Fulke Greville records that his Majesty replied to this 'with Emotion & Agitation, "You have quitted a profession I have always loved, & You have Embraced one I most heartily detest"'.[3] After this meeting, Willis told the King that 'his ideas were now deranged' and that he would require Willis's constant attention and 'management'.[4] Around the same time, the Prince of Wales wrote to his brother that their father had suffered 'a total loss of all rationality' and was now 'a compleat lunatick'.[5] The arrival of Willis, his son (also a physician) and their entourage of assistants into the midst of the royal household at Kew and, especially, the other royal physicians, led to weeks of jealousies, conflicts and contested diagnoses and treatments. Willis insisted on complete authority over who had access to the King, and relied upon methods of physical restraint (sometimes using a restraining-chair, sometimes a strait-waistcoat), combined with a quasi-mesmeric technique of commanding the King's submission with his gaze. (This episode has been well documented by historians of medicine and dramatized in Alan Bennett's play, the Madness of the George III and the film based on it.)[6]

The controversy about the efficacy and propriety of Willis's regime was at its height in January 1789, when a period of regency seemed likely to become a necessity. For six days, a Parliamentary Select Committee interviewed Willis and the other physicians, to try to ascertain the prospects, if any, of his Majesty recovering sufficiently to retake the reins of government in the near future. The exchanges were published by order of Parliament, as well as in various commercial editions, and in the daily press.[7] Almost the entirety of *The Times* for 16 January, for example, was taken up with extended extracts from 'The Examination of the King's Physicians', focussing especially on what Willis himself had to say. When challenged on his policy of excluding others from seeing the King without his express permission, Willis replied that the unexpected appearance of physicians or family members might 'excite troublesome emotions' and that such troublesome emotions might 'retard the cure of the patient'. In answer to another question, Willis spoke of the great benefit his Majesty had enjoyed from being allowed, at the appropriate time, to see his wife and daughters. To allow him to continue in 'anxiety' at his separation from his family would have been harmful, Willis said. Even a brief glimpse of his daughters had had the effect of 'softening him into tears'; the King 'shewed the greatest marks of parental affection I ever saw'. The committee asked whether 'observing those emotions, which may naturally take place at the sight of relations or friends' provided the doctor with any basis for judging on the likelihood of a cure in this case. Willis replied that it was a very favourable symptom for a patient to show 'affection rather than aversion' in such circumstances. Willis made much of the King's recently recovered ability to read literature and make intelligent comments upon what he had read (although he was taken to task by the committee for having allowed the King, inadvertently it seems, to get his hands on a copy of the uncomfortably apposite *King Lear*). All parties to the discussion agreed that recovery would be indicated by further signs that 'the understanding is strengthened' and that his Majesty's 'frequent gusts of passion' (brought on by contradiction or irritation) were becoming less frequent and shorter in duration.[8] Subsequently, the King made a surprisingly rapid recovery, to the extent that the regency, which had seemed inevitable, was forestalled. On 19 March, the Lord Mayor of London attended his Majesty at Kew and, speaking on behalf of the 'Aldermen and Commons of the City of London', begged leave to approach the throne with 'the most heart-felt emotions of joy and exultation', and gratitude that Divine Providence had restored the sovereign to health. The King thanked the Lord Mayor for 'this fresh mark of your loyalty, and of your affection for my person'.[9] Writing to the King in May 1789, Dr Willis commented that the King's illness had 'drawn out the affections of millions of people' who wished to see him restored to health.[10]

This famous episode, then, provides the historian with a valuable insight into the way that the languages of passion, affection and emotion were deployed in a highly charged medical and political arena in England, on the eve of the French Revolution. It also poses a series of questions, and alerts us to a set of problems of interpretation, which I examine below. These concern the meanings of some key terms. The historian should hesitate, for instance, before assuming that 'emotion' in this context meant what it means today, or even what it meant in the nineteenth century. The best interpretation in some cases might simply be 'movement', 'agitation' or 'disturbance'; in others it might be more like the modern psychological sense of 'feeling'; in others it possibly refers to physical expressions, such as the King's tears.[11] It is notable that Dr Willis was concerned about 'troublesome emotions', but encouraged familial 'affection', and considered the King's 'gusts of passion' (meaning, primarily, uncontrolled anger) to be among the leading and most worrying symptoms of the disorder, which was admitted by all to be a disorder of the 'intellect' or 'understanding'. The drama at Kew provides not only a vivid example of the relationship between emotions, health and disease, but also an illustration of the centrality of moral and political authority to discussions of management of the insane. That a mad-doctor was, eventually, allowed to command and physically restrain a King against his will must tell the historian something about the attitudes of the culture that allowed it, and attitudes towards reason, passions and insanity as well as towards power. This episode occurred not only on the eve of the French Revolution but also on the eve of epochal changes in the treatment of the insane; with manacles, straitjackets and other physical restraints, along with medicinal cures, increasingly being replaced by moral and emotional forms of management – the process described by Foucault as the 'birth of the asylum'.[12]

Words, categories, disciplines

Of the many different ways one might go about investigating the histories of medicine and emotion, perhaps the most ambitious would be to attempt to recover somehow the actual emotional experiences of historical actors (such as King George, his family and his physicians). My purpose in this chapter is more modest. I aim to document and analyse some shifts in the languages of medicine and emotion. My particular focus is on the shift that took place between the late eighteenth and mid-nineteenth centuries from theorising about 'passions' and 'affections' to theorising about 'emotions'. I am interested in the meanings of the theoretical terms historians use when trying to make sense of the mental experiences of those they study, and also in the terms employed by those people themselves. The principal point that I want to make is straightforward: that it is important to be aware of differences between the categories employed by medical practitioners and their patients in the eighteenth and nineteenth centuries and those employed by twenty-first-century historians. As the historian of psychology Roger Smith has recently argued, there is a reflexive relationship between the psychological categories used to understand the human mind and mental experience itself. An understanding of the history of those categories is, therefore, an essential part of any attempt to understand the experiences they were used to define, describe, evoke or control.[13]

My initial interest in the history of affective categories arose from the realisation that historians of ideas had tended to assume that theories of the 'passions' in the pre- and early-modern periods could be treated as theories about what modern thinkers called 'emotions', without anything significant being lost in that translation. I began to wonder whether such a shift in words should really be treated as merely terminological. In the resulting study, *From Passions to Emotions*, I have suggested that just as historians of chemistry would not think it satisfactory to consider Joseph Priestley's theory of dephlogisticated air a theory of 'oxygen', so historians of the human sciences would benefit from trying to avoid similar anachronism in the case of important psychological categories. Theories of 'passions' and theories of 'emotions' were not generally theories of the same things. The verbal change signified theoretical, social and cultural changes. Those two key terms were not defined the same way, nor did they refer to the same set of mental phenomena, and nor, perhaps most importantly, did they have the same connotations and associations. They occupied different, albeit overlapping, semantic fields.[14]

This statement needs to be qualified. I do not mean to imply that either 'passion' or 'emotion' has had a singular or entirely stable meaning during this or any other period. There have been various theories of 'passions' and of 'emotions'; these categories have been employed as parts of quite various moral, religious, medical and psychological theories and practices. So, for instance, Descartes thought the passions of the soul were produced by the movements of the animal spirits and their impact on the brain; for David Hume, the passions were produced by mental impressions in combination with ideas.[15] Thomas Brown described emotions, in his 1820 *Lectures on the Philosophy of the Human Mind*, as vivid feelings with mental causes, while William James, in his 1884 article entitled 'What Is an Emotion?' famously answered that it was a vivid feeling of visceral bodily movements.[16] It is therefore possible to make the same point about two different theories of the 'passions', or two different theories of the 'emotions', as I wish to make more generally about theories of 'passions' on the one hand and of 'emotions' on the other: they do not necessarily refer to the same things or rest on the same assumptions about body and mind. What I particularly wish to draw attention to, however, is the way that, in English-language works by philosophers, physicians and psychologists, the well-established category of the 'passions' was, along with related categories, gradually displaced by the 'emotions' as the leading theoretical term during the nineteenth century. Although no two authors precisely agreed about the extent and meaning of either category, the 'passions' (and 'affections') generally had stronger religious, moral and medical connotations, and often belonged to a discourse about the soul (in its relationship with the body) and about various kinds of pathology. 'Emotion', in the eighteenth century, had various meanings, the most common being, as I have already noted, a generalized kind of agitation or disturbance. By the end of the nineteenth century, however, it was an established theoretical term closely associated with the emerging discipline of psychology, and was often discussed with reference to somatic, evolutionary and neurological mechanisms. Although the 'passions' had often been a very broad and undifferentiated category, it had also been combined and contrasted by many writers with other categories such as 'appetites', 'affections' and 'sentiments'. The possession of moral sentiments and religious affections, along with the power of reason, marked humanity's superiority over the brutes (which were motivated purely by appetites and passions). From the outset 'emotions' tended to be used even more broadly and with fewer such distinctions.[17]

This study is an example of what Raymond Williams, in his *Keywords*, called 'historical semantics'.[18] The attempt to pursue history through a study of the changing meanings of English words goes back at least to the 1850s, and the work of the Church of Ireland clergyman and philologist Richard Chenevix Trench; this approach was taken further in the twentieth century by two friends, and founding members of the Oxford 'Inklings', Owen Barfield and C. S. Lewis.[19] One of the functions of such histories was, Lewis wrote in 1959, 'to aid more accurate reading' of past texts.[20] We might today think of our goal as being to read past texts in ways that are historically sensitive and contextually informed, rather than 'accurate'. Nonetheless studies such as Lewis's certainly constitute a valuable aid to careful historical reading.

The history of words can, and should, however, be more than 'merely lexical' (a phrase that Lewis used, with undue and misleading modesty, to describe his own *Studies in Words*).[21] As Eric Hobsbawm put it in the opening sentence of *The Age of Revolution*: 'Words are witnesses which often speak louder than documents'; witnesses, that is, to historical changes that extend far beyond the 'merely lexical'.[22] One of my hopes in this chapter is that changes in the uses of the words 'passion', 'affection' and 'emotion' will serve as witnesses, not only to new kinds of medical and psychological theorising in the nineteenth century, but also to professional, institutional and, particularly, disciplinary changes. The picture of British and American intellectual life that I sketch below is, unsurprisingly no doubt, one of increasing professional specialisation and intellectual segregation; one in which connections between psychology, medicine, morality and religion gradually seem to have become harder and harder to make. Teachers and practitioners of medicine around the turn of the nineteenth century composed works in which moral, philosophical and religious teachings about the passions were interwoven with information about organic diseases and medical treatments. During the second half of the nineteenth century, while theories of 'the emotions' were coming to the fore, the earlier, more multifaceted, and highly moralized approach to passions and affections was starting to fracture into several less closely connected discussions, each relating to an emerging specialist academic discipline. Philosophers and psychologists produced expert treatises on the emotions of the normally functioning mind; medical psychologists specialized in theories of 'moral insanity' and 'emotional insanity'; and religious and theological concerns were left to an increasingly embattled and isolated clergy.

Physicians and metaphysicians in Edinburgh

Writing in 1905, James Mark Baldwin noted that 'The use of the word emotion in English psychology is comparatively modern. It is found in Hume, but even he speaks generally rather of passions or affections'. Baldwin went on to remark that 'When the word emotion did become current its application was very wide, covering all possible varieties of feeling, except those that are purely sensational in their origin'.[23] The individual whom I have previously described as the 'inventor of the emotions' was the Edinburgh physician and philosopher Dr Thomas Brown (1778–1820).[24] It was in Brown's very widely read *Lectures on the Philosophy of the Human Mind*, published shortly after his death, that 'emotion' was first systematically used as the term for all those vivid mental states previously understood, variously, as appetites, passions, affections and sentiments. The president of Yale, Noah Porter, later remarked that 'The influence of Brown's terminology and of his methods and conclusions has been potent in the formation and consolidation of the Associational Psychology – represented by J. Mill, J. S. Mill, Alexander Bain, and Herbert Spencer'.[25] The physician and psychologist Henry Holland, who had studied medicine in Edinburgh in the early years of the century, agreed that Brown's popular *Lectures* contained 'much that has been appropriated, doubtless unconsciously, by later writers on Mental Philosophy'.[26] It is not my intention, however, to rehearse here the details of Brown's importance in the history of philosophy, nor the broader influence of his work on later nineteenth-century psychological theorising.[27] Instead, I want to look at the transition from passions to emotions that Brown's lectures helped to effect from the perspective of the history of medicine. Almost all the early works to adopt the general psychological category of 'the emotions', from the 1830s to the 1850s, were written either by philosophical writers indebted to Brown, or by physicians. In the former category were works by Friedrich Rauch, George Ramsay and William Lyall[28] in the latter were works by Thomas Burgess, William Cooke and Walter Johnson.[29]

Brown himself qualified as a physician in Edinburgh, taking his MD in 1803, and making a name for himself as the 'coadjutor' of the controversial and flamboyant James Gregory, Professor of the Practice of Medicine, and Edinburgh's leading consulting physician. According to Brown's biographer, however, he found medicine 'a laborious profession, whose frequent and agitating interruptions were ... unfavourable to close and continuous thought'.[30] He did not hesitate, therefore, when the opportunity arose, to relieve an ailing Dugald Stewart of the duty of delivering the University's lectures in moral philosophy (thereby ensuring that his intellectual activities were henceforth less frequently disturbed by patients or passions). Brown was appointed to the chair in moral philosophy, jointly with Stewart, in 1810, and held the post until he died in 1820, at the age of 42. Moral philosophy was not the only Edinburgh chair for which Brown had been considered. He had previously been an unsuccessful candidate for the chairs of rhetoric (in 1801) and logic (in 1808).[31] In addition to his abilities in the fields of medicine and philosophy, Brown was a productive poet (his collected poetical works filled four volumes).[32] This ability to move between various literary and intellectual worlds was not altogether unusual. To take just one other example, Erasmus Darwin, with whom Brown engaged in a lively philosophical debate in print in the late 1790s, had similarly combined philosophical, medical and literary pursuits.[33] As L. Stephen Jacyna has put it, the 'weakness of the barriers between different forms of discourse' was 'a leading feature of British intellectual life at the turn of the nineteenth century'.[34] As we shall see below, medicine, morals and metaphysics were intertwined in various ways during this period.

That is not to say, however, that there were no boundaries whatever between these areas of eighteenth-century culture. To stay with the example of Brown himself, his MD thesis, *De Somno*, his philosophical *Lectures* and his published poetry were all clearly distinguishable by such things as content, language, genre and audience. These were productions relating to different parts of the intellectual and cultural life of the time. There is no question that Brown and his contemporaries saw some clear distinctions and boundaries between what are now considered as different areas of 'discourse'. James Gregory, for instance, was said to have described Brown's 'poetry as too philosophical, and his philosophy as too poetical'.[35] For him, at least, then, these two kinds of writing could and should have been kept more separate.

I have claimed that Brown's category of the 'emotions' was part of a discourse of the mind (with its own history) that tended to neglect theological and religious concerns about troubling appetites and passions, religious and domestic affections and moral and aesthetic sentiments. One might expect to find that theories of the 'emotions', which developed from the lectures on mental philosophy produced by an Edinburgh physician such as Brown would be engaged with medical understandings of the human mind and body, even if not with moral and theological ones. The suggestion I wish to make in this chapter, however, is that this was not entirely the case. Rather, the theoretical transition from 'passions' and 'affections' to 'emotions' was one which saw the practical and moral advice of medical writers, as well as the theological concerns of earlier theorists of the soul, being gradually pared away to leave a more purely psychological account of the properly functioning mind. Rick Rylance has made a connected, more general point: that mainstream Victorian psychology frequently failed to engage with or integrate ideas about human mental debility and pathology. Rylance describes this as the result of a separation between psychology and psychiatry, corresponding to a black- and-white division between the sane and the insane.[36] In addition to the Victorian desire to distinguish the normal mind very clearly from the abnormal, there was another separation developing, between general medicine and medical psychology, which Michael Donnelly has documented.[37] In pre-1850 medical discussions of passions and emotions, healthy minds were not so clearly separated from unhealthy ones, and bodily and mental aspects of disease were more often kept simultaneously in view (even though all the authors I consider below clearly stated their belief that the passions and emotions were states of a separate soul or mind, which interacted with the physical brain and body). Despite the increasingly physiological tenor of scientific psychology, then, the detachment of that psychology from both general medicine and from psychiatry in some measure obscured connections between body and mind in both health and disease, connections which the passions and emotions of the mind had clearly revealed to earlier medical writers.

The period from 1789 to around 1850 was, then, one during which the treatment of and theorising about the insane was being transformed by the widespread introduction of asylums and regimes of 'moral treatment'. At the same time, physiology, psychology and psychiatry were emerging as distinct, and controversial, scientific specialisms; and anxieties about the French Revolution (and about kings losing their heads), materialistic philosophy and the operation of ungoverned human passions cast a long and menacing shadow over discussions of the relationship between mind and body. Finally, it was a period when many still sought to keep scientific and philosophical speculation within a respectable natural-theological framework, in which mental adaptations, as well as physical ones, were evidence of the power, wisdom and goodness of a benevolent deity. The texts produced in this period that I discuss below are united by language and subject. All are English-language texts. They include medical dictionaries and handbooks, volumes of lectures delivered by professors of medicine to their students and philosophical and psychological treatises written by physicians. They are further united by their connection with a single city, namely Edinburgh. Edinburgh's eighteenth-century reputation as the 'Athens of the North' was based, of course, not only on its excellence in philosophy, law and the natural sciences, but also on its status as one of the leading centres in Europe for the teaching of the theory and practice of medicine, under such luminaries as William Cullen and John Gregory. The authors of virtually all the sources discussed below were connected with the world of Edinburgh medicine one way or another. Most, including William Falconer, Benjamin Rush, Thomas Brown, Charles Bell, James Cowles Prichard, Thomas Burgess and William Carpenter graduated as doctors of medicine in Edinburgh between the 1760s and 1830s[38]; others, such as William Cullen, John Gregory and James Gregory, were themselves professors of medicine in Edinburgh[39]; others, like the physician-philosopher Anthony Florian Madinger Willich and the authors of various medical dictionaries, had a less direct connection, but composed works that were used by students in Edinburgh during this period.

What I hope to recover, then, are some of the ways that theories and practices connected with medicine (as well as philosophy and religion) in Edinburgh in the late eighteenth and early nineteenth centuries shaped how people thought and wrote about passions, affections and emotions as the nineteenth century wore on. I have identified three interconnected areas of discourse through which to undertake this survey. In the first section, on 'pathologies and affections', I look at the various ways that physicians understood the influence of the passions of the soul upon the patient's bodily frame, whether as causes or cures of organic disease. Under the heading of 'the importance of government', I look at the ways that physicians and others were called upon to master their passions (and to control their patients), and the connection between these medical imperatives and political anxieties in this age of revolution. In the third section, on 'madness and morals', I try to map out a complicated semantic field in which physicians wrote about the 'moral' causes, consequences and treatments of insanity. Finally, I reflect upon how, and with what consequences, a scientific psychology of emotions emerged from this milieu.

Pathologies and affections

In 1787 the Medical Society of London instituted the Fothergillian Medal, which was to be awarded, in the first year of its existence, to the author of the best dissertation written on the following question: 'What diseases may be mitigated or cured, by exciting particular affections or passions of the mind?' The winning entry was judged to be that submitted by William Falconer, MD, who had studied at both Leyden and Edinburgh, and whose dissertation was published in 1788 as *A Dissertation on the Influence of the Passions upon Disorders of the Body*.[40] The phrasing of the Medical Society's question is itself interesting, using as it does a distinctive phrase of the period, namely 'the affections or passions of the mind'. 'Affection' and 'passions' each received a substantial entry, with several different senses, in Samuel Johnson's 1755 *Dictionary of the English Language*.[41] Each could be used as a very general term, and the two could, in some contexts, be used interchangeably. However, each also had its own specific connotations. Passions were often more violent and troubling. Affections, on the other hand, like moral sentiments and feelings of sympathy, were generally more benevolent and virtuous inclinations. Both 'passion' and 'affection' also had particular medical meanings. *The Edinburgh Medical and Physical Dictionary*, published in 1807, and advertised as having been compiled 'agreeably to the doctrines of Cullen, Monro, Hunter, Fordyce, Gregory, Denham, Saunders, Home, and other modern teachers in Edinburgh and London', defined an 'affection' as 'any disorder of the whole body, or a part of it, as in the hysterics, cholic, & c.' Thus, the entry continued, 'by adding a descriptive epithet to the term *affection*, most distempers may be expressed. We say *febrile affection, cutaneous affection*, & c.'[42] The entry for the term 'passion' in the same dictionary illustrates well the permeability of boundaries between different areas of discourse in this period. The entry made reference to the opinions of Cicero, of the Scottish philosopher Thomas Reid and of one Dr Willich, the author of a series of lectures on diet and regimen. The entry also defined the passions in terms of the Galenic distinction between 'naturals' and 'non-naturals' (the former were those things that naturally belonged to an animal, the latter included air, meat, drink, sleep, motion, rest and excretion). 'In medicine', the entry continued, 'the passions make one of the non-naturals, and produce very sensible effects on the body. *Joy, anger, love*, and *fear*, or *grief*, are the principal'. The rest of the entry was an almost verbatim quotation from Dr Willich's lectures, to which I shall return shortly, from a section dealing with both moral and medical aspects of the passions. Other related entries in the dictionary included those for 'caeliac passion', or diarrhoea, and 'hysteric passion', another term for hysteria. Both 'affection' and 'passion' were described as being expressed in Greek as *pathos* (meaning something undergone, something suffered or a passion). Two further technical terms with their roots in the Greek *pathos* were included in the dictionary: '*pathetici*' and '*pathologia*'. The former was a set of nerves, so named 'because they direct the eyes to express the passions of the mind'; the latter, of course, was simply the Latin term for '*pathology*; or the doctrine of diseases'.[43] In 1826, *The New London Medical and Surgical Dictionary* reproduced many of these definitions, including the one for 'affection' (but omitting the long, more philosophical entry on the passions).[44] The new edition of Robert Hooper's *Lexicon Medicum; or Medical Dictionary*, in 1848, included very similar definitions;

'affection' could be a term for disease, 'thus an *inflammatory affection*, a *nervous affection*'; or for a condition of the mind, 'thus we speak of the *moral affections*'. Hooper's *Lexicon* (first published in 1798), as with earlier medical dictionaries, had no entry for 'emotion', but under the heading of '*pathemata animi*' offered an account of the effects of the exciting or depressing passions of the mind on the body.[45] Again, in common with several other dictionaries, the ideas in the entry for the passions agreed in several respects with the account offered in Willich's lectures.

Willich, like Erasmus Darwin, Thomas Brown, John Gregory and James Gregory, combined the roles of physician and metaphysician.[46] His *Lectures on Diet and Regimen*, which seems to have been a widely consulted source of information on medicine and the passions in this period, was published in 1799.[47] A year earlier, his *Elements of the Critical Philosophy* had also been published in London. This work, one of the first accounts of Kant's philosophy to be made available to the English-speaking world, was an enthusiastic summary of Kant's ideas. The first more critical appraisal of Kant to appear in English was included in the second number of the *Edinburgh Review*, five years later. Its author was another philosophical physician – Thomas Brown.[48] Although Willich and Brown were both involved, then, in the overlapping worlds of medicine and philosophy at the turn of the nineteenth century, they had very different approaches to the passions and emotions of the mind; largely, no doubt because of differences of context and audience. Brown's lectures were philosophical rather than physiological; they were concerned with the science of the mind *qua* mind; it made little difference to Brown what particular physical processes were involved; he sought to establish the laws linking one mental process to another, which was what he meant by 'mental physiology'. Willich's lectures, on the other hand, were intended to be of more practical use, were aimed at a wider audience and constantly considered how the pathology and physiology of the body were affected by the passions of the mind.[49] The content of Willich's lectures was both medical and moral. Passions, he wrote, 'operate upon the body either suddenly, or slowly and gradually. Sudden death, or imminent danger of life, may be the consequence of the former: a gradual decline and consumption, that of the latter'. All passions, whether agreeable or disagreeable, he warned could be harmful, indeed fatal, if they rose 'to a high and violent degree'. Terror could bring on a fatal apoplexy; anxiety, an epileptic fit. The 'affections' of joy and hope, on the other hand, in moderation, could enliven the bodily frame and 'contribute to the preservation of health and life, more than all the medicines that can be contrived'. In most cases, however, the passions, Willich thought, tended to produce dangerous and irregular motions in the body. *'Hence the dominion over our passions and affections'*, he wrote, in an italicized passage that would later be used in the *Edinburgh Medical and Physical Dictionary*, *'is an essential and indispensable requisite to health'*.[50] Willich combined medical science with moral sentiment, warning that, among 'the mournful passions', for instance, an 'extravagant degree of love' was one of the most harmful passions. Such a passion, Willich wrote, not only 'transgresses the bounds of reason', but is apt to disorder the understanding and consume 'all the vital powers, by a slow fever'.[51] In conclusion, Willich urged his readers to 'combine solid principles of virtue' with a 'sober and active life' in order to repress the violent passions, 'these enemies to human life'.[52]

Other medical writers of the period echoed these ideas about the need to resist and restrain the passions for the good of one's mental, moral and physical health, on the one hand, and about their potential use in the relief of diseases on the other. This applied to William Cullen's treatment of the subject in his *First Lines of the Practice of Physic*, and to the appendix to the 1829 edition of that work, written by James C. Gregory (son of James Gregory, the Edinburgh Professor of the Practice of Medicine). It was also true of several other works, including those by the Edinburgh-trained American physician (and signatory of the Declaration of Independence) Benjamin Rush.[53] The passions of the mind, these authors all agreed, were physically dangerous. As Rush summarized it, 'consumptions, fevers, convulsions, disease of the stomach and bowels, visceral obstructions, apoplexy, palsy, madness, with a numerous and melancholy train of mental disease, are all frequently brought on by the undue actions of the passions upon the body'.[54] It was also a commonplace that extreme passions, whether of grief, terror or even joy could be fatal. William Carpenter, in his *Principles of Human Physiology*, recounted a case where an infant died after being breast-fed by a woman in a state of excessive emotional excitement (as a result of intervening in a violent quarrel between her husband and their lodger).[55] A mother's passions could be fatal for her child as well as herself. 'Emotions', too, would later be cited as potential threats to female health. An American physician, writing on abortion in the 1860s, listed 'emotional causes' among the leading etiological factors (along with physical and medicinal causes): 'Women have aborted immediately after hearing of the death of a beloved husband; or the gnawing canker of grief, shame, and remorse, has led to loss of the foetus; and martyred women have aborted at the stake'.[56]

On the other hand, the passions could also sometimes be used themselves as cures for bodily diseases. Slow fevers could sometimes be cured by a paroxysm of anger, Rush told his students; terror could cure convulsions and rheumatism; and joyful laughter had been known to cure children of rickets. Rush included an account of a cardinal in Rome who was cured from his sickness by laughing as he watched 'a favourite pet monkey put on his pontifical robes, and strut about his bedchamber, with the solemn face with which he had often seen his master perform his public ecclesiastic duties'.[57] And it was not only agreeable passions that could serve as remedies for physical ailments; Rush also described a case of a woman cured by her avarice (when all else had failed, she was restored to health when some coins were placed in her hand) and of an 'instance related by Dr Darwin, of a man who revived himself, when debilitated by exercise of labour, by thinking of a person whom he hated'. All this, Rush concluded, was 'striking proof of the efficacy of the malignant passions in imparting vigour and strength to the body, when affected by debility, or a disease of weak and morbid action'.[58]

The importance of government

In the late eighteenth century, the languages of passions and politics were frequently mixed. There were discussions, as we have seen, of how the passions could cure the body, and, less frequently, of how a greater role for the common people might improve the health of the body politic. More frequently, however, the analogy was between controlling the passions and controlling the people. As Roy Porter has noted, this analogy between rational government of the people and rational government of the passions had been an irresistible one ever since Plato wrote on the subject.[59] In this time of insane Kings and violent revolutions, it took on a new urgency. Edmund Burke had been one of the advocates of a swift move to a regency in December 1788, when the King's mental condition showed little sign of improvement. During a heated Parliamentary debate on the state of the King's health, Burke, urging coolness and moderation, said that 'if ever there was a time for moderation, and a Government of the passions it was the present'. It was reported, however, that Burke had failed to follow his own advice on this score and had flown into 'a great passion' and alleged that William Pitt was competing with the Prince of Wales for the role of Regent.[60] Burke's position seemed to be that a king who could not govern his own passions was unfit to govern his people. Here, as in his famous *Reflections on the Revolution in France*, published two years later,

Burke juxtaposed political and psychological pathologies. In the *Reflections*, Burke argued that government was not made 'in virtue of natural rights', as radicals such as Thomas Paine would have it, but as a 'contrivance of human wisdom' to satisfy certain human wants. Among these wants, Burke argued, was the need 'not only that the passions of individuals should be subjected, but that even in the mass and body as well as in the individuals, the inclinations of men should frequently be thwarted, their will controlled, and their passions brought into subjection'. That was his attitude to the passions of the 'swinish multitude'. When it suited him, however, Burke would subvert this traditional subjection of the passions to reason and recommend quite the reverse. While contemplating recent political events, he wrote, it was right and natural that 'our passions instruct our reason'; specifically the passion of pity, felt by right-minded people, for the dethroned and abused French king. Such moral feelings should determine the intellect to reject the political doctrines of the revolution. In this instance, Burke wrote, 'We are alarmed into reflection'.[61] It seemed that, for Burke, the question of whether passions should be allowed to instruct one's reason or needed to be governed by it depended very much on the class, ideology and sympathies of the individual in question.

Benjamin Rush, a political man himself, made an explicit connection in his lectures between the usefulness of knowledge of the passions to the physician *qua* physician, and the usefulness of this knowledge to him as 'a citizen and a man'. In the world of public affairs, Rush said, a physician's knowledge of the debilitating effects of confinement on the understanding and the passions would make him 'an enemy to slavery and a friend to liberty' and from his knowledge of the destructive effects of the passions upon the human body he would support only governments 'which filter laws most completely from the passions of legislators, judges, and the people'.[62] In medical and moral treatises, even when political concerns such as Burke's and Rush's were not mentioned, individuals were encouraged to repress and master their own passions through the use of reason and will, for the sake of their health as well as for the sake of civilized society.

Similar images of mastery and government arose in two other contexts connected specifically to the practice of medicine. The doctor had to have mastery both over his patients and over his passions (both 'patients' and 'passions', like 'pathology', of course, had the same etymology in the Greek *pathos*). John Gregory's *Lectures on the Duties and Qualifications of a Physician*, first published in 1772, was still a standard text when it came out in its final version in 1805, in an edition revised and corrected by his son, James Gregory. Historians have debated the extent to which these lectures constituted a founding text in the history of medical ethics.[63] My particular interest here is in how John Gregory addressed the physician's relationship both with his passions and with his patients. In the case of the former, he lamented the fact that physicians had 'often been reproached with hardness of heart' as a result of their apparent ability to deal with human misery while remaining entirely composed. To be genuinely insensible would be a 'great misfortune to a physician' since it would deprive him of 'one of the most natural and powerful incitements to exert himself for the relief of his patient'. On the other hand, excessive passions of anxiety or sympathy might cloud a physician's understanding, depress his spirits and 'prevent him from acting with that steadiness and vigour, upon which perhaps the life of his patient in a great measure depends'.[64] The physician, then, needed to steer a middle path, keeping at all times in command of his passions, but without repressing entirely his sympathy and sensibility. Physicians should 'feel whatever is amiable in pity, without suffering it to enervate or unman them'.[65] Very similar arguments had been made in relation to religion in the eighteenth century, by writers such as Isaac Watts, who had argued that true religion lay between the extremes of cold rationalism and overheated enthusiasm. Watts had written that he hoped 'neither to turn religion into a matter of speculation or cold reasoning, nor to give up the devout Christian to all wandering fooleries of warm and ungoverned passion'.[66] The physician and the minister, in the overlapping spheres of medical and religious practice, each offered the same moral advice to their charges: to govern their passions, but while still fostering their feelings of sympathy and human affection.

One man who singularly failed to follow this advice was John Gregory's son James. Despite the fact that James had brought out an edition of his father's celebrated lectures on the duties of the physician, he was almost constantly in dispute with his professional colleagues in Edinburgh throughout his own career as a medical practitioner. One source of controversy was his appointment of Thomas Brown as a 'coadjutor', an arrangement whereby patients consulting Gregory by letter would, if he were away from Edinburgh or too pressed to reply himself, receive a reply from Brown. Most famous among Gregory's many disputes was his 1793 quarrel with the Professor of Midwifery, James Hamilton. It culminated in Gregory beating Hamilton with a cane. Having been tried and found guilty of this offence, and sentenced to pay a fine of £100, Gregory was reputed to have responded that he would happily pay twice that amount for the opportunity to do it again.[67] Gregory also fell out with Andrew Duncan, Professor of the Institutes of Medicine, and Senior Censor of the Royal College of Physicians of Edinburgh. In an open letter of 1811 (which seems to have arisen from an accusation made by Gregory that Duncan had stolen an unprinted work of his), Duncan attacked Gregory for a range of offences, including the appointment of Brown as a partner, and several matters concerning rules for membership of the College itself and the dispensing of medicines by its members. In the rash, intemperate and scandalous ways that Gregory had conducted himself in all these disputes, Duncan concluded, it had become clear that his professions of openness and candour were nothing but 'hypocritical quackery' and that, in sum, he was 'a wrong-headed man, who allowed precipitate passion, and a suspicious temper, to overcome his judgment' – a thoroughly unworthy son for the late John Gregory, author of the 'excellent work on the Duties and Qualifications of a Physician'.[68] There was no doubt in Duncan's mind that in failing to govern his passions James Gregory had shown himself to be an incapable medical practitioner.

Finally, the physician had to command and govern badly behaved patients, who, like his passions, had the potential to 'ruffle his temper, and consequently to cloud his judgment, and make him forget propriety and decency of behaviour'. In order to ensure that unruly patients follow the prescribed diet and medicines, John Gregory had said, 'the government of a physician over his patient should undoubtedly be absolute' (although in reality, he acknowledged, compromises would have to be made).[69] Patients and their pathologies needed to be governed by composed and rational physicians just as the physician's own passions needed to be moderated and controlled by a clear intellect and settled will. It was just these questions about government and mastery that had come to a head so uncomfortably in Francis Willis's shocking determination to defeat the King's disease by morally mastering George III himself.

Madness and morals

Today the word 'moral' and cognate terms refer almost always to some part of the realm of ethics. In the eighteenth and nineteenth centuries, they had a much broader range of meanings. Jeremy Bentham, for instance, wrote, of his 1789 *Introduction to the Principles of Morals and Legislation*, that as 'an introduction to the principles of *morals*' it contained matters pertaining to 'the extensive ideas signified by the terms *pleasure, pain, motive* and *disposition*'; in addition, if he had wished to produce a more complete study, Bentham wrote, he should also 'have given a similar analysis of the not less extensive though much less determinate ideas annexed by the terms *emotion, passions, appetite, virtue, vice* and some others'.[70] As this list indicates, the sphere of 'morals' included a broad range of mental feelings and faculties, whether directly connected to questions of right and wrong or not. The various connotations of the term can be brought out more precisely by considering the terms with which it was contrasted. 'Moral' could be used to suggest a contrast with 'intellectual'; the passions and emotions of the mind could thus be distinguished from its rational operations, as 'active' or 'moral' rather than intellectual powers. William Cooke, MD, delivering an address to the Hunterian Society in 1839, on the subject of *Mind and the Emotions Considered in Relation to Health and Disease,* used the term 'moral feelings' as a synonym for emotions, which could be distinguished from the mind's 'intellectual faculties'.[71] 'Moral' could also be opposed to 'physical'. The pursuit of 'moral science', for instance, which for some simply meant ethics, for others meant the scientific study of human character as opposed to the human body. So, for example, *The Phrenological Journal* was renamed after 1837 *The Phrenological Journal and Magazine of Moral Science*. The sixth book of John Stuart Mill's *System of Logic* (1843) was an account of 'The Logic of the Moral Sciences'; and in 1851 Cambridge University introduced two new triposes, one in the 'natural sciences', the other in the 'moral sciences'.[72] In these cases 'moral sciences' indicated systematic knowledge in a broad range of areas, including philosophy, history, political economy and the sciences of man and society; and this was contrasted with the scientific study of the physical world.

How, then, was this multivalent language of 'morals' deployed in medical discussions of deranged passions and morbid emotions in this period? Physicians discussed moral causes of insanity, moral forms of insanity, and moral treatments of insanity. Starting with causes, James C. Gregory, for instance in his 1829 appendix to Cullen's *First Lines,* divided the 'remote and predisposing causes to insanity' into the physical and the moral.[73] Among the physical causes, Gregory wrote, hereditary predisposition was the most significant. The moral causes of insanity, he continued, 'are numerous and their influence is very extensive. Among them may be included all the passions and emotions of the mind, all impressions which tend to excite or exalt the imagination, particularly those caused by great political events and commotions, and new or abstruse views upon doctrinal points in religion'. In support of this statement, Gregory went on to cite the analysis made by Jean-Etienne Dominique Esquirol of the most common moral causes of insanity among patients admitted to the Salpêtrière in Paris. These included 'domestic vexations', 'misery and reverses of fortune', 'thwarted love', 'political events', 'jealousy', 'excessive study' and 'disappointed ambition.' Gregory also speculated that the 'exciting emotions' such as joy had more frequently brought on insanity than the 'depressing emotions of sorrow or disappointment'. Gregory reported a case related by his late father, Professor James Gregory, of a mother and her two daughters who 'all became insane in one day, immediately after a sudden accession of fortune'.[74]

The sexual passion was singled out for particular mention as a source of insanity by some medical writers. In his 1858 *Dictionary of Practical Medicine*, James Copland devoted a whole paragraph in his preface to denouncing the 'secret vice' of 'self-pollution', in both sexes, as the leading cause (among many) responsible for the 'increasing prevalence of insanity' among the young. Copland felt that the 'great importance of guarding against this most vile and degrading vice' had not been 'sufficiently warned against by medical writers'. More had to be done, he wrote, to prevent the spread of this 'most destructive physical and moral evil'.[75] In a footnote to his *Principles of Human Physiology*, first published in 1839, William B. Carpenter had taken a similar moral tone in warning young physicians against the physical and mental damage that could be done by 'illicit gratification of the sexual passion'. Carpenter commented that some might feel such considerations 'misplaced in a Physiological Treatise', yet he felt sure that he would not be blamed for dealing with such an important subject as the potential 'melancholy consequences of the violation of the Divine law in this particular'.[76]

Passions and emotions, then, were among the leading 'moral causes' of insanity, as well as of physical disease. They were also fundamental to one particular form of insanity. First distinguished by the Edinburgh-trained physician James Cowles Prichard in his 1835 *Treatise on Insanity*, as 'moral insanity', this condition, of disordered emotions combined with an apparently sound intellect, later became known as 'emotional insanity' (and later still as 'psychopathy').[77] Insanity had generally been understood, as it was at the time of King George's mental illness in the 1780s, as an intellectual deficit. Theorists following Locke believed that insanity was not a deficiency in the powers of reason, *per se*, but a form of disordered reasoning which made valid deductions from false premises, such as the premise that one was made of glass or that one's friends were planning one's destruction (or that one was engaged in a love-affair with Lady Pembroke).[78] William Cullen had similarly described the various forms of insanity as instances of reasoning based on one particular mistaken judgement.[79] Copland's 1858 dictionary said that the two leading forms of insanity, 'maniacal or raving insanity' and 'incoherent or fatuous insanity', were both primarily disorders of the intellect but that 'some recent writers' had added a new category of 'moral insanity; in which the intellectual powers are but little, or not very manifestly, impaired; the disorder appearing chiefly in the emotions, habits and conduct'.[80] This new category in one way expanded the sense of the pathology of the passions and emotions by marking them out as, along with the intellect, potential seats of insanity. John Bucknill even tried to reverse the traditional assumption, arguing that insanity was always, in the first instance, an emotional rather than an intellectual disorder.[81] Discussions of moral or emotional insanity certainly thus spread the net of psychiatry much wider by removing the condition that a person must suffer cognitive impairments such as delusions in order to qualify as insane.

In another way, though, the category of 'moral insanity' represented a strengthening of boundaries, both between intellect and emotion (each could now separately be judged healthy or diseased), and between normal and abnormal emotional states (disordered emotions implied 'moral insanity'; normal ones need not be considered pathological). In connection with the former distinction, the category was also objected to by many, in legal and ethical contexts, on the grounds that it could be used to recategorize certain acts as cases of insanity rather than of crime, sin or vice. In 1845, for instance, *The Times* reported on a case concerning a disputed will. One party to the dispute had alleged that the deceased had been afflicted with 'moral insanity'. The court heard from medical witnesses, including William Jenner, who 'exposed what he considered the absurdity of the new doctrine', which was attributed to Pinel and Prichard. The court was satisfied that the doctrine could not be safely applied to legal questions: 'If a man who was free from delusions (as the deceased in this case was) and capable of acts of business (as he was), might nevertheless be held to have been insane, it would involve this branch of testamentary law in utter confusion'. The conclusion was that it was safer 'to rely upon the ancient and general doctrine of these courts', that there was no insanity without intellectual delusions.[82] Two years later, the category of 'moral maniac' was debated again in connection with the sensational case of twelve-year-old William Allnutt, who was charged with murdering his wealthy grandfather with a dose of arsenic. On this occasion, *The Times*, in a leading article, condemned the suggestion that the boy was suffering from a mental disease affecting his moral sense. The leader writer wrote that 'Moral insanity is merely another word for crime', and should be recognized as such. The jury agreed and found the boy guilty, rejecting his defence of insanity. However, they also 'earnestly recommended the prisoner to mercy on account of his tender age'. The judge rejoiced that the jury had 'thrown to the winds the idle sophistry' of the plea of moral insanity, and found that Allnutt was indeed guilty of 'the crime of wilful murder'. However, he agreed that his age presented grounds for mercy, and so recommended that his young life be spared, although he should 'pass the whole of the remainder of it in ignominy and disgrace'.[83]

In addition to discussing moral causes and moral forms of insanity, this period saw physicians introducing 'moral treatment' of the insane in their asylums. This was a régime that replaced the more physically repressive approach favoured by mad-doctors of the previous generation, such as Francis Willis, as well as their uses of practices such as blistering and bleeding. Benjamin Rush, whose views on the passions' effects on the body were examined above, also wrote about their effects on the mind. Indeed he has been described as the 'father of American psychiatry', and his 1812 *Medical Inquiries and Observations upon the Diseases of the Mind* as 'the first American book on psychiatry'.[84] In the chapter of that work on the 'Derangement of the Passions', Rush considered each of the passions in turn, explaining how it had been implanted in the human mind for wise and benevolent purposes but could still, in certain circumstances, become deranged. The remedies for deranged passions were divided into physical, rational and moral. Suggested remedies for the fear of death, for example, included inculcating in the patient sound 'opinions of the divine government and the relation we sustain to the great Author of our being'; reassuring a patient that death was not always accompanied by pains; keeping them distracted and talking about other things; listening to music; or giving them opium.[85] A similar cornucopia of treatments was proposed for each of the other cases of excessive or disturbed passions; mixing physical remedies such as alcoholic spirits and laudanum with rational and moral ones, including reasoning, distraction, persuasion, reading of scripture and religious teaching.[86] Copland's *Dictionary* also included correct religious instruction (in distinct contrast to mistaken or fanatical religious views), along with music, other 'employments and amusements', and the visits of friends, as being among the most effective kinds of 'moral management' of the insane.[87]

Perhaps the most famous example of the 'moral treatment' of insanity in this period was that developed at the Quaker asylum in York known as the Retreat, founded in 1796 by Samuel Tuke. Anne Digby's account of 'moral treatment' at the Retreat and elsewhere in this period shows how this phrase encompassed several of the senses of 'moral' that have already been discussed. The phrase implied 'a concentration on the rational and emotional rather than the organic causes of insanity', and 'was not so much a specific technique as a range of non-medical treatments designed to involve the patient actively in his recovery'.[88] Moral management, then, involved engaging the sympathies and affections of the insane individual as a stimulus to the restoration of their reason, rather than relying purely upon physical restraint and the administering of medicines. Foucault has offered a suspicious account of the real motives behind the establishment of the apparently liberal and humane Retreat. This account emphasizes the desire of the Society of Friends to exercise control over members who had lost their reason, both by keeping them out of large public establishments, whose profane inmates might corrupt them, and by keeping them in a state of religious and moral anxiety: 'Tuke created an asylum where he substituted for the free terror of madness the stifling anguish of responsibility'.[89] On either a charitable or a suspicious reading, however, this new kind of management was certainly a 'moral' régime. As such, in seeking to restore reason through manipulation and nurture of the patient's affective life, it represented the psychiatric application of Hume's famous philosophical dictum that 'Reason is, and ought only to be the slave of the passions, and can never pretend to any office than to serve and obey them'.[90]

'Moral' then could, finally, imply a contrast with 'medical' as well as with 'intellectual' or 'physical' or 'natural'. If we had to translate 'moral' into one or more modern-day terms, it would very often carry both 'psychological' (more specifically 'emotional') and 'ethical' connotations. A moral treatment could, like a psychosomatic cure, be one that was a result of the action of the mind on the body. But, similarly, it could be one that was humane and respectful (and possibly religiously motivated), as opposed to unethically repressive or intrusive. Indeed, the replacement of the language of 'moral' treatments and 'moral' disorders by terms associated with the discourses of psychology and psychiatry, on the one hand, and philosophy and ethics, on the other, may be a good example of how the history of language can shed light on deeper cultural movements; in this case suggesting a divorce between medical and moral understandings of the mind from the middle of the nineteenth century onwards.

Psychology's demoralized emotions

How, then, did the new category of 'emotion' fit into this context? As we have seen, the term had various physical and psychological meanings prior to the nineteenth century, but lacked the theoretical medical connotations of 'passion' and 'affection'. This is not to say that physicians and their patients did not use the term 'emotion'; they certainly did, as we have seen. However, there are, to my knowledge, no entries for 'emotion' or cognate terms in any medical dictionary prior to 1890, when George Gould's *New Medical Dictionary* contained an entry for 'emotional' and a sub-entry for 'emotional insanity'.[91] Noah Webster's *American Dictionary of the English Language*, in 1828, offered eight meanings each for 'affection' and 'passion', and three for 'emotion'.[92] The two authorities cited to illustrate the meaning of the latter were the Scottish writer Lord Kames, who had made a distinction in the middle of the eighteenth century between passions (which involved desires) and emotions (which did not), and the physician, philosopher and dissenting English theologian Thomas Cogan, who had suggested, in several works on the passions in the early years of the nineteenth century, that emotions were the sensible and visible effects that passions of the mind produced on the bodily frame.[93] Several examples from the 1780s and 1790s confirm that Cogan was not alone in this usage. A *Times* reporter wrote, in 1785, about an unscrupulous young man who, in the midst of an elaborately fraudulent romantic suit (which involved, among other things, pretending to be the Prince of Wales), was said to have 'feigned those involuntary emotions which indicate a rising passion'; a similar sense may have been intended in the discussion of the King's 'emotions' on seeing his family, discussed at the outset; and it is certainly evident in Jeremy Bentham's assertion in his 1789 *Introduction to the Principles of Morals and Legislation* that 'The emotions of the body are received, and with reason, as probable indications of the temperature of the mind'.[94] Willich's usage in 1799 also seems to have been similar to Cogan's. Willich described laughter (rather than joy, which was a passion of the mind) as 'a salutary emotion' undergone by the body.[95] Similarly, Rush, in his anecdote about the cardinal and his monkey, spoke of laughter as a 'gentle and pleasing emotion of the diaphragm and its contiguous muscles' connected with the passion of joy.[96]

The processes by which religiously and morally engaged discussions of passions and affections gave way to a more detached psychology of emotion in the nineteenth century, and ultimately to a frequently uncritical celebration of feeling and emotion in contemporary culture, were not straightforward. A sense of the potential moral pathology of the emotions was certainly present in Thomas Brown's *Lectures*, in which he warned his students that although among the emotions were to be found 'those tender regards which make our remembrances sacred' and the 'delightful wonder' that arises from the contemplation of works of art and nature, they also included a much more terrible class of mental affections, 'the enemies with which our moral combat, in the warfare of life, is to be carried on'. This formidable class of emotions included 'every passion which can render life guilty and miserable, – a single hour of which, if that hour be an hour of uncontrolled dominion, may destroy happiness for ever, and leave little more of virtue than is necessary for giving all its horror to remorse'.[97]

So, although Brown, the main instigator of the shift to the category of 'emotions', was himself trained in the medical tradition I have been discussing and clearly shared in the moral sensibilities of that tradition, later psychological writers picking up his new over-arching category of 'emotions' along with his desire to reduce knowledge of the mind to a set of strictly scientific laws, showed a tendency to neglect the pathology of the passions. Instead, they wrote about emotions as they occurred in the normal, healthy mind in a way that was primarily concerned with careful introspective observation on the one hand, and an understanding of the functions of the brain and the nervous system on the other. Psychological textbooks, such as two by Scottish psychologists, Alexander Bain's *The Emotions and the Will* (1859) and James McCosh's *The Emotions* (1880), combined introspective mental science with neurophysiological findings, but had much less to say about medical, mental, moral or even political pathology.[98]

Histories of the emotions often point to the English naturalist Charles Darwin's 1872 *Expression of the Emotions in Man and Animals*, and the American philosopher–psychologist William James's 1884 *Mind* article 'What Is an Emotion?' as the founding texts of our modern scientific understandings of the emotions. I hope this chapter has helped to substantiate the idea that these works by no means represent the beginning of the story. Darwin, of course, had, like his grandfather, studied medicine in Edinburgh, albeit abortively, and his theory of the expression of the emotions was closely intertwined with the medico-theological accounts of two Edinburgh MDs, Charles Bell and Thomas Burgess.[99] William James had trained as a physician and, in the philosophical reading of his youth had been, by his own account, 'immersed in Dugald Stewart and Thomas Brown'.[100] Historians of the philosophy and psychology of the emotions, then, as well as historians of medicine and psychiatry, may have something to learn by looking back, beyond imposing figures of the later nineteenth century, such as Darwin, James and Freud, to university lecture halls, hospitals and asylums in Britain and America in the decades following the French Revolution.

Just as some ancient writers had thought of the passions as *morbi*, or diseases of the soul,[101] so physicians and philosophers in this period understood the passions as potentially pathological moral states that were intimately connected with the affections of the body. Their theories were based on a mixture of physiological science, medical anecdote, practical experience, trial and error, religion and moralism. The remedies they suggested were psychosomatic, even holistic, simultaneously addressing hygiene, diet, exercise, occupation, morals and beliefs. When psychology emerged as a discipline distinct from medicine and moral philosophy later in the nineteenth century, its theories of the emotions were undoubtedly based on improved understandings of neurophysiology and on vastly more rigorous experimental investigations. However, the strengthening of boundaries between philosophical, moral, medical and psychiatric specialisms meant that these theories of emotion less frequently engaged with moral and therapeutic dimensions. Those were just the dimensions that had come together in earlier attempts by physicians to understand and to control their patients and their passions.

Notes

1. I am very grateful to Janet Browne and Fay Bound Alberti for inviting me to contribute to the Wellcome Trust 'Medicine and Emotion' workshop in October 2004 from which this volume arose; and to them and the other participants in that workshop for helpful suggestions and challenging questions. I am also grateful to Emily Butterworth and Tim Hickman for their thoughts and advice on this chapter.
2. From the diary of Queen Charlotte's lady-in-waiting, Fanny Burney, later Madame D'Arblay, quoted in Ida Macalpine and Richard Hunter, *George III and the Mad-Business* (London: Allen Lane, 1969), p. 52.
3. Quoted in ibid., p. 53.
4. Robert Fulke Greville's diary, quoted in ibid, p. 53.
5. Quoted in ibid., p. 55.
6. Macalpine and Hunter, *George III,* especially Chapters 3 and 4; Roy Porter, *A Social History of Madness: Stories of the Insane* (London: Phoenix, 1996), Chapter 3; Alan Bennett, *The Madness of George III* (London: Faber, 1992); *The Madness of King George* dir Nicholas Hynter (Samuel Goldwyn and Channel Four Films, 1994).
7. Macalpine and Hunter, *George III,* pp. 75–6.
8. *The Times,* 16 January 1789, especially pp. 3–4.

9. Reported in *The Times*, 23 March 1789, p. 2.
10. Macalpine and Hunter, *George III*, p. 95.
11. On the various meanings of 'emotion' prior to the nineteenth century, see Thomas Dixon, *From Passions to Emotions: The Creation of a Secular Psychological Category* (Cambridge: Cambridge University Press, 2003), pp. 62–6, 104–9.
12. Michel Foucault, *Madness and Civilisation: A History of Insanity in the Age of Reason,* trans. R. Howard (London: Routledge, 2001 [1961]), Chapter 9.
13. Roger Smith, 'The History of Psychological Categories', *Studies in History and Philosophy of Biological and Biomedical Sciences* (2005), 36, pp. 55–94; on 'emotion', see pp. 85–9.
14. Dixon, *From Passions to Emotions,* especially pp. 13–19.
15. Ibid., especially pp. 75–9, 104–9.
16. Ibid.; Thomas Brown, *Lectures on the Philosophy of the Human Mind* (Edinburgh: Tait, 1828), reprinted in *Life and Collected Works of Thomas Brown 1778–1820,* ed. Thomas Dixon, 8 vols. (Bristol: Thoemmes Press, 2003), vol. 6; William James, 'What Is an Emotion?', *Mind* (1884), 9, pp. 188–205.
17. Dixon, *From Passions to Emotions,* especially pp. 20–5.
18. Raymond Williams, *Keywords: A Vocabulary of Culture and Society* (London: Croom Helm, 1976), p. 11.
19. Richard Chenevix Trench, *On the Study of Words* (London: Parker, 1851); idem, *English Past and Present* (London: Parker, 1855); Owen Barfield, *History in English Words* (London: Methuen, 1926); C. S. Lewis, *Studies in Words*, 2nd ed. (Cambridge: Cambridge University Press, 1967).
20. Lewis, *Studies in Words*, p. vii.
21. Ibid.
22. Eric Hobsbawm, *The Age of Revolution: Europe 1789–1848* (London: Cardinal, 1973), p. 13.
23. James Mark Baldwin (ed.), *Dictionary of Philosophy and Psychology,* 3 vols. (London: Macmillan, 1905), vol. I, p. 316.
24. Dixon, From Passions to Emotions, pp. 109–34.
25. Noah Porter, 'Philosophy in Great Britain and America: A Supplementary Sketch', in Friedrich Ueberweg (ed.), *History of Philosophy from Thales to the Present Time,* Vol. II. *History of Modern Philosophy* (London: Hodder and Stoughton, 1874), pp. 349–460, 410.
26. Henry Holland, *Recollections of Past Life* (London: Longmans, Green and Co., 1872), p. 85.
27. These are subjects I have explored in 'Introduction', to Dixon (ed.) *Thomas Brown,* vol. I, pp. v–xxxi; Dixon, *From Passions to Emotions,* pp. 109–34.
28. Friedrich Rauch, *Psychology: Or, a View of the Human Soul, Including Anthropology,* 2nd ed. (Delmar: Scholars' Facsimiles and Reprints, 1975 [1841]); George Ramsay, *Analysis and Theory of the Emotions* (London: Longman, Brown, Green and Longmans, 1848); William Lyall, *Intellect, the Emotions, and the Moral Nature* (Edinburgh: Constable, 1855). See Thomas Dixon, 'George Ramsay', in W. J. Mander and Alan Philip Frederick Sell (eds), *Dictionary of Nineteenth-Century British Philosophers,* 2 vols. (Bristol: Thoemmes Press, 2002); idem, 'William Lyall', in ibid.
29. William Cooke, *Mind and the Emotions Considered in Relation to Health and Disease* (London: Longman and Co., 1839); idem, *A Commentary of Medical*

and Moral Life; or Mind and the Emotions, Considered in Relation to Health, Disease, and Religion (London: Longman, Brown and Co., 1852); Thomas H. Burgess, *The Physiology or Mechanism of Blushing; Illustrative of the Influence of Mental Emotion on the Capillary Circulation* (London: Churchill, 1839); Walter Johnson, *The Morbid Emotions of Women; Their Origin, Tendencies, and Treatment* (London: Simpkin, Marshall, and Co., 1850). On Burgess, see Dixon, *From Passions to Emotions*, pp. 164, 170–1. For some basic biographical information on Cooke, see Frederic Boase, *Modern English Biography*, 6 vols. (Truro: Netherton and Worth, 1892–1921). The first of Cooke's works was the first English-language book to use 'the emotions' in the title in its new psychological sense. In the much expanded 1852 edition, Cooke refers readers who 'desire to pursue this subject closely and philosophically' to the 'interesting works of Drs Brown and Abercrombie' (p. 114), that is to say Thomas Brown and John Abercrombie. On Abercrombie, see Thomas Dixon, 'John Abercrombie', in Mander and Sell (eds), *Dictionary*.

30. David Welsh, *Account of the Life and Writings of Thomas Brown MD* (Edinburgh, Tait, 1825), reprinted in Dixon (ed.) *Thomas Brown,* vol. I, p. 169.
31. Ibid., pp. 163–7.
32. Thomas Brown, *The Poetical Works of the Late Thomas Brown,* 4 vols. (Edinburgh: Tait, 1820).
33. On Brown's critique of Darwin's *Zoonomia,* see Dixon (ed.) *Thomas Brown,* vol. I, pp. vii–x; Welsh, *Account,* pp. 42–71.
34. L. Stephen Jacyna, 'Immanence or Transcendence: Theories of Life and Organization in Britain, 1790–1835', *Isis* (1983), 74, pp. 310–29, 312.
35. Henry Cockburn, *Memorials of His Time* (Edinburgh: Foulis, 1909), p. 347.
36. Rick Rylance, *Victorian Psychology and British Culture 1850–1880* (New York and Oxford: Oxford University Press, 2000), pp. 113–4.
37. Michael Donnelly, *Managing the Mind: A Study of Medical Psychology in Early Nineteenth-Century Britain* (London and New York: Tavistock Publications, 1983), p. 103.
38. William Falconer received an Edinburgh MD in 1766 and one from Leyden the following year. See *ODNB*. Benjamin Rush started his medical training in Philadelphia, travelled to study under William Cullen in 1766, and took his MD in Edinburgh in 1768. See Lisbeth Haakonssen, *Medicine and Morals in the Enlightenment: John Gregory, Thomas Percival and Benjamin Rush* (Amsterdam and Atlanta, GA: Rodopi, 1997), pp. 190–5. Thomas Brown and Charles Bell were both in James Gregory's class around the turn of the century, Brown receiving his MD in 1803; see 'List of Students (Practice of Medicine)', Edinburgh University Library, Da 35 PRAC. MED. 1; Dixon, (ed.) *Thomas Browne* vol. I, p. vii; Welsh, *Account,* pp. 83–9. James Cowles Prichard received his MD from Edinburgh in 1808, his thesis forming the basis of his later work on the history and varieties of the human race. See *ODNB*. Thomas Burgess's work on *The Physiology or Mechanism of Blushing* was published in 1839 and advertized the author as an MD and graduate of the University of Edinburgh. William B. Carpenter studied under William Pulteney Alison and took his MD in 1839. See *ODNB*.

39. William Cullen was Professor of Medicine and Chemistry from 1755 to 1766, Professor of the Institutes of Medicine from then until 1773, and then Professor of the Practice of Medicine until 1790. John Gregory was Professor of the Practice of Medicine from 1766 until his death in 1773. His son, James Gregory was Professor of the Institutes of Medicine from 1776 until 1790, and from then until he died in 1821 was Professor of the Practice of Medicine. On the Gregories see James Fieser, 'Introduction', in James Gregory, *Philosophical and Literary Essays*, 3 vols. (Bristol: Thoemmes Press, 2001), vol. I; Agnes Grainger Stewart, *The Academic Gregories* (Edinburgh and London: Oliphant, Anderson and Ferrier, 1901). For more on the structure, appointments processes, and incumbents of Edinburgh chairs in this period, see J. B. Morrell, 'The University of Edinburgh in the Late Eighteenth Century: Its Scientific Eminence and Academic Structure', *Isis* (1971), 62, pp.158–71.
40. William Falconer, *A Dissertation on the Influence of the Passions upon Disorders of the Body* (London: C. Dilly, 1788). See Richard Hunter and Ida Macalpine (eds), *Three Hundred Years of Psychiatry, 1535–1860: A History Presented in Selected English Texts* (London: Oxford University Press, 1963), pp. 507, 592–3.
41. For more on eighteenth-century uses, see Dixon, *From Passions to Emotions,* pp. 62–97. On Johnson's analyses of the passions and its relationship to eighteenth-century and more ancient faculty psychologies, see Alan T. McKenzie, 'The Systematic Scrutiny of Passion in Johnson's Rambler', *Eighteenth–Century Studies* (1986–1987), 20, pp.129–52.
42. Robert Morris and James Kendrick, *The Edinburgh Medical and Physical Dictionary,* 2 vols. (Edinburgh and London: Bell and Bradfute; Mondell, Doig and Stevenson, 1807).
43. Ibid.
44. J. S. Forsyth, *The New London Medical and Surgical Dictionary* (London: Sherwood, Gilbert and Piper, 1826).
45. Robert Hooper, *A Compendious Medical Dictionary* (London: Murray and Highley, 1798); idem, *Lexicon Medicum; or Medical Dictionary*, 8th ed. (London: Longman, Brown, Green, and Longmans, 1848).
46. John Gregory was, with his cousin Thomas Reid, one of the founding members of the Aberdeen philosophical society, and was Professor of Philosophy in Aberdeen before becoming Professor of Medicine in Edinburgh. See Haakonssen, *Medicine and Morals*, especially pp. 19–35, which brings out the many close interconnections between the worlds of medicine and philosophy in the eighteenth century; Laurence B. McCullough, *John Gregory and the Invention of Professional Medical Ethics and the Profession of Medicine,* (Dordrecht: Kluwer, 1998). On James Gregory, see also Stewart, *The Academic Gregories* and Fieser, 'Introduction', in Gregory, *Philosophical and Literary Essays*.
47. Anthony Florian Madinger Willich, *Lectures on Diet and Regimen*, 2nd ed. (London: Longman and Rees, 1799).
48. See René Wellek, *Immanuel Kant in England, 1793–1838* (Princeton: Princeton University Press, 1931), pp. 7–15, 32–8. Thomas Brown, 'Villers' Kant', *Edinburgh Review* (1803), 1, pp.253–80, reprinted in Dixon (ed.) *Thomas Brown,* vol. 3.

49. It is interesting to compare the awareness of physicians in this period of how their management of a patient's passions could alter the course of their disease with the account offered of the clinical management of the body through emotions in the late nineteenth and early twentieth century by Otniel B. Dror. See 'Creating the Emotional Body: Confusions, Possibilities, and Knowledge', in P. N. Stearns and J. Lewis (eds), *An Emotional History of the United States* (New York and London: New York University Press, 1998), pp. 173–94, especially pp. 178–80.
50. Willich, *Lectures on Diet and Regimen*, pp. 577–9.
51. Ibid., p. 589.
52. Ibid., p. 603.
53. See William Cullen, *First Lines of the Practice of Physic*, new ed., with an appendix by James Craufurd Gregory, 2 vols. (Edinburgh and London: Bell and Bradfute and Adam Black; Longman, Rees, Orme, Brown and Green, 1829); Haakonssen, *Medicine and Morals*; Benjamin Rush, *Benjamin Rush's Lectures on the Mind*, ed. Eric T. Carlson, Jeffery L. Wollock, and Patricia S. Noel (Philadelphia: American Philosophical Society, 1981); idem, *Sixteen Introductory Lectures, to a Course of Lectures upon the Institutes and Practice of Medicine* (Philadelphia: Bradford and Innskeep, 1811); idem, *Medical Inquiries and Observations, upon the Diseases of the Mind* (Philadelphia: Kimber and Richardson, 1812).
54. Rush, *Benjamin Rush's Lectures*, p. 259.
55. William B. Carpenter, *Principles of Human Physiology, with their Chief Applications to Pathology, Hygiène, and Forensic Medicine, Especially Designed for the Use of Students*, 2nd ed. (London: John Churchill and Sons, 1844), pp. 361–2.
56. Edwin M. Hale, *A Systematic Treatise on Abortion* (Chicago: C. S. Halsey, 1866), p. 45.
57. Rush, *Benjamin Rush's Lectures*, p. 262.
58. Ibid., pp. 263–4.
59. Porter, *Social History of Madness*, pp. 39–59.
60. *The Times*, 11 December 1788, p. 2.
61. Edmund Burke, *Reflections on the French Revolution, and Other Essays* (London: Dent, 1910), pp. 57, 76–8.
62. Rush, *Benjamin Rush's Lectures*, pp. 269–70.
63. John Gregory, *Lectures on the Duties and Qualifications of a Physician*, revised and corrected by James Gregory, MD (Edinburgh and London: William Creech; T. Cadell and W. Davies, 1805). For discussions of these lectures and the history of medical ethics, see Haakonssen, *Medicine and Morals* and McCullough, *John Gregory*.
64. Gregory, *Lectures*, pp. 8–9.
65. Ibid., p.19.
66. Isaac Watts, *Discourses of the Love of God, and its Influence on All the Passions; With a Discovery of the Right Use and Abuse of Them in Matters of Religion*, 3rd ed. (London: Oswald and Buckland, 1746), p. xi. On the religious and theological dimensions of discussions of passions and affections in this period, see Dixon, *From Passions to Emotions*, pp. 62–97. For a study of the connections between Methodist religion and medicine in the work of George Cheyne, see David E. Shuttleton, 'Methodism and Dr George Cheyne's

"more enlightening principles"', in Roy Porter (ed.), *Medicine in the Enlightenment* (Amsterdam and Atlanta, GA: Rodopi, 1995), pp. 336–59. On similar concerns about enthusiasm in the context of nineteenth-century Quaker contributions to the treatment of the insane, see Fiona Godlee, 'Aspects of Non-Conformity: Quakers and the Lunatic Fringe', in William F. Bynum, Roy Porter, and Michael Shepherd (eds), *The Anatomy of Madness: Essays in the History of Psychiatry*, 3 vols. (London and New York: Tavistock Publications, 1985), vol. 2, pp. 73–85.

67. *ODNB*.
68. Andrew Duncan, *A Letter to Dr James Gregory of Edinburgh* (Edinburgh: A. Constable and Co., 1811), pp. 9, 12, 22, 32.
69. Gregory, *Lectures*, pp. 17, 21.
70. Jeremy Bentham, *An Introduction to the Principles of Morals and Legislation*, eds J. H. Burns and H. L. A. Hart, with a new introduction by F. Rosen (Oxford: Clarendon Press, 1996 [1823]), p. 3.
71. Cooke, *Mind and the Emotions*, pp. 2, 8, 20, 38; see also Cooke, *A Commentary*, pp. vii, 114. On Cooke's natural theology of the emotions, see Dixon, *From Passions to Emotions*, pp. 189–90. On the strong separation between intellect and emotions in later nineteenth-century psychology, see ibid., pp. 227–9.
72. John Stuart Mill, *A System of Logic, Ratiocinative and Inductive, Being a Connected View of the Principles of Evidence, and the Methods of Scientific Investigation*, 2 vols. (London: Parker, 1843).
73. Cullen, *First Lines*, p. 387.
74. Ibid., pp. 391–3.
75. James Copland, *A Dictionary of Practical Medicine*, 3 vols. (London: Longman, Green, Longmans, and Roberts, 1858), vol. 1, pp. xxii–xxiii. For an extensive history of medical and moral attitudes to masturbation in the eighteenth and nineteenth centuries, see Thomas Laqueur, *Solitary Sex: A Cultural History of Masturbation*, (New York: Zone Books, 2003).
76. Carpenter, *Principles*, p. 360n.
77. Hunter and Macalpine (eds), *Three Hundred Years*, pp. 836–42; Dixon, *From Passions to Emotions*, p. 228; Vieda Skultans, *Madness and Morals: Ideas on Insanity in the Nineteenth Century* (London, Routledge and Kegan Paul, 1975), Chapter 6; idem, *English Madness: Ideas on Insanity 1580–1890* (London: Routledge and Kegan Paul, 1979), pp. 65–8.
78. The latter, of course, was one of King George's delusions; see Macalpine and Hunter, *George III*, Chapter 4; Porter, *Social History of Madness*, p. 50.
79. Copland, *Dictionary*, vol. 2, p. 433.
80. Ibid.
81. John C. Bucknill, 'Correspondence on the Theory of Emotional Insanity', *Journal of Mental Science* (1874), 20, pp. 484–6.
82. *The Times*, 1 November 1845, p. 8.
83. *The Times*, 16 December 1847, pp. 7–8; 24 December 1847, pp. 4–5.
84. Eric T. Carlson, Jeffrey L. Wollock, and Patricia S. Noel, 'Introduction', to Rush, *Benjamin Rush's Lectures*, pp. 1–43, p. 2.
85. Rush, Benjamin Rush's Lectures, pp. 324–9.
86. Ibid., pp. 314–46

87. Copland, *Dictionary,* vol. 2, pp. 535–7.
88. Anne Digby, 'Moral Treatment at the Retreat, 1796–1846', in Bynum, Porter, and Shepherd (eds), *The Anatomy of Madness,* vol. 2, pp. 52–72, on p. 53. See also idem, *Madness, Morality and Medicine: A Study of the York Retreat, 1796–1914,* (Cambridge: Cambridge University Press, 1985); Roy Porter, 'Shaping Psychiatric Knowledge: The Role of the Asylum', in idem (ed.), *Medicine in the Enlightenment,* pp. 256–73; Donnelly, *Managing the Mind;* Godlee, 'Aspects of Non-Conformity'.
89. Foucault, *Madness and Civilization,* Chapter 9, p. 234.
90. David Hume, *A Treatise of Human Nature,* ed. L. Selby-Bigge and P. Nidditch (Oxford: Clarendon, 1978), p. 415; Dixon, *From Passions to Emotions,* p. 106.
91. George M. Gould, *A New Medical Dictionary* (London: H. K. Lewis, 1890).
92. Noah Webster, *An American Dictionary of the English Language* (New York, Converse, 1828).
93. On Kames, see Adela Pinch, *Strange Fits of Passion: Epistemologies of Emotion, Hume to Austen* (Stanford: Stanford University Press, 1996), pp. 4–6, 49–50. Cogan, who received his MD from Leyden for a dissertation on the passions, wrote three books on the subject, which were widely cited in the first half of the nineteenth century: Thomas Cogan, *An Ethical Treatise on the Passions* (Bath: Hazard and Binns, 1807); idem, *A Philosophical Treatise on the Passions* (Bath: S. Hazard, 1802); idem, *Theological Disquisitions; or an Enquiry into those Principles of Religion, which are Most Influential in Directing and Regulating the Passions and Affections of the Mind* (Bath: Hazard and Binns, 1812).
94. *The Times,* 2 June 1785, p. 3; Bentham, *Introduction,* p. 63.
95. Willich, *Lectures on Diet and Regimen,* pp. 581–2.
96. Rush, *Benjamin Rush's Lectures,* p. 262.
97. Brown, *Lectures,* p. 339.
98. Alexander Bain, *The Emotions and the Will* (London: Parker, 1859); James McCosh, *The Emotions* (London: Macmillan, 1880).
99. Dixon, *From Passions to Emotions,* pp. 159–79.
100. Quoted in ibid., p. 204.
101. On classical views of the passions, and medieval Christian appropriations and corrections of such views, see ibid., pp. 26–61.

3

Languages and Landscapes of Emotion: Motherhood and Puerperal Insanity in the Nineteenth Century[1]

Hilary Marland

The pathologization of emotions: The emergence of puerperal insanity

Motherhood was deemed to be an emotional experience in the nineteenth century in manifold ways. Bearing and raising children was held up as women's sacrosanct duty, and the fulfilment of this was imbued with great emotional content. It was anticipated that childbirth would result in joy and contentment, and cement households and families in loving and dutiful relationships. In well-to-do families, motherhood, formerly an ascribed function associated with producing heirs, was reconfigured as an 'achieved status', which called upon the 'emotional, intellectual and moral talents' of women.[2] By the late eighteenth century, William Buchan was exhorting mothers to take full responsibility for raising their children: 'Nothing can be more preposterous than a mother who thinks it below her to take care of her own child, or who is so ignorant as not to know what is proper to be done for it.'[3] In the words of the influential German physician Christian Struve, which were translated into English in 1801, 'The greatest charms and dignity of a woman are derived from her maternal office, and a good mother equally deserves the affection of her husband and the esteem of the world.'[4] Motherhood was to move to the apex of the Victorian bourgeois ideal: 'virtually any reference to motherhood in the social texts of the era seemed to call forth, as if by necessity, yet one more recitation of the maternal creed. We hear endlessly of the mother's sacred mission to rear children'[5] Yet the act of becoming a mother was also recognized and represented as a time of extreme suffering and danger. The death of Princess Charlotte in childbirth in 1817, following a 50-hour labour and massive haemorrhaging, was to become emblematic of the extreme hazards of motherhood, coalescing medical anxieties as well as igniting public fears about the possible tragedies awaiting women in childbed.[6] Early nineteenth-century medical texts laid out in great detail a wide spectrum of disorders, accidents, diseases and physical ailments associated with pregnancy and birth. As well as having the potential to result in great bodily harm and even death, childbirth was described increasingly as likely to result in mental ruin.

Puerperal insanity entered the medical canon in the early nineteenth century, as a term used to describe the mental breakdown of women following childbirth. It was characterized by a deep and persistent depression or, more frequently, violent mania. It was related to the sister disorders of insanity of pregnancy and lactation, and both psychiatric and obstetric practitioners busied themselves describing the symptoms and claiming expertise to treat it. While doctors were sanguine about patients' chances of recovery compared with other forms of mental disorder, with around three-quarters of cases being declared cured or significantly improved after several months of treatment, it was troubling in its prevalence, tendency to reoccur and for the fact that some women did not recover, lapsing into permanent melancholia or dementia.[7] Puerperal mania in particular stunned observers for its sudden appearance so soon after a woman had become a mother and for the ferocity of its symptoms:

> There is a change in temper, an irascibility inducing snappish remarks, and a peculiar hurried manner; sleep is much disturbed, the countenance betrays distrust, the pulse is rapid, and the patients are generally voluble. They often become negligent of their infant. At length both behaviour and language are incoherent, and delirium is fully developed. Acts of violence, sometimes suicide, are in this stage committed before the nature of the malady is suspected.[8]

In a similar vein, James Reid reported 20 years later, in 1848, how

> talking gradually becomes almost incessant, and generally on one particular subject, such as imaginary wrongs done to her by her dearest friends; a total negligence of, and often strong aversion to her child and husband are evinced; explosions of anger occur, with vociferous and violent gesticulations; and although the patient may have been remarkable previously for her correct, modest demeanour, and attention to her religious duties, most awful oaths and imprecations are uttered, and language used which astonishes her friends.[9]

Melancholia, though less flamboyant in its appearance, was feared just as much, if not more, for its insidiousness and intractability; it crept rather than stormed into the bosom of the family, leading often to a lapse into permanent insanity.

Prior to the nineteenth century, pregnant women and young mothers were recognized as having the potential to fall prey to mental disorders, and to display odd, disturbing, and sometimes aggressive behaviour. The physician Charles Lepois noted in the mid seventeenth century that some women developed mental disorders during childbirth or delirium immediately after;[10] more alarmingly, women destroying their newborn children while suffering 'temporary phrenzy' were reported in courtrooms and by the press.[11] Childbirth was seen as a challenging experience, not only as a rite of passage but also as a time marked by fear of harm or even death, a dread expressed by women in diaries and letters as their confinements approached. Anne Hulton, pregnant herself, saw her friend die in childbirth: 'it did much affect me with cares and thoughts about another world, which had been too little minded by me'.[12] One mother prophesied in 1801, 'My dearest child, When this is delivered unto you, the hand that writes it will be mouldered in dust.'[13] Midwife Martha Mears, offering a more measured account of the challenges to women approaching confinement, cautioned of the need to remove the dread and terrors felt by many pregnant women, and advised how knowledge of nature's ways, sensible words and care and tenderness would 'expel the dark phantoms of the brain'.[14]

When the term 'puerperal insanity' entered medical literature and public consciousness in the early nineteenth century it carried with it new meanings and content. While the sad and gloomy or erratic, crazed and disruptive behaviour of women around the time of childbirth had far from passed unnoticed before the nineteenth century, it was also deemed to be a 'natural' part, albeit an unpleasant and deeply troubling part, of the process of pregnancy and birth. It also tended to be described in terms of individual incidents rather than as a collective phenomenon likely to affect large numbers of women, and was dealt with as it occurred rather than being anticipated. John Woodward reported in 1716 on an extraordinary case of a woman who became obsessed with the notion that her child had been harmed by her seeing a porpoise swimming in the River Thames, a sight that had at first much delighted her.[15] After delivery the woman became suicidal and was tempted to 'fling her Child into the Fire, beat its Brains out, and the like; to which she had the utmost Horror and Aversion . . . She had frequent Temptations to lay violent Hands on herself'.[16] Following heavy purging by Woodward, the woman recovered. At no time did Woodward refer to the woman's distressed state as being a 'disease' or a general marker of women's propensity to fall prey to such influences. He linked it to the influence of maternal imaginings rather than any pathological state of mind or body.

Puerperal insanity was strongly rooted in a number of changes that took place in medicine and psychiatry around the turn of the nineteenth century. Bolstered by an outpouring of writing on midwifery and the diseases of women – the two becoming closely bound together – childbirth was rebranded during the early decades of the century as an event that was brim full of danger. An entire battery of diseases and disorders associated with childbirth or following hard on the heels of delivery were elaborated on, with textbooks and advice books listing in grim detail preternatural labours, with breech, footling, knee or transverse presentations; labours complicated by distortion of the pelvis; swelling of the 'soft parts'; convulsions; retention of the placenta; uterine haemorrhage; disorders of the breast, womb or lochia; prolapse or inversion of the uterus; rupture; puerperal convulsions; phrenitis and fever; fibroids and cancers. By 1844 Robert Lee was offering little in the way of optimism in his *Lectures on the Theory and Practice of Midwifery*, and we can only hope that few mothers-to-be ever read his introduction:

> There are many diseases peculiar to women; they are all exposed to great suffering and danger during pregnancy and child-bearing, and many die from acute disorders following delivery. Not only are the functions of the uterine system before conception often disturbed, but all the different parts of this system undergo morbid alterations of structure, which are frequently of a malignant nature, and excite, in their progress towards a fatal termination, the most acute and protracted pain.[17]

The turn of the nineteenth century was also marked by a new professional paradigm with increasing numbers of women being delivered by men-midwives and obstetricians, a process that went hand in hand with a shift from associating expertise in childbirth with communities of women to self-proclaimed new male specialists. Fears connected to birth up until the nineteenth century were largely contained within family clusters and groups of women, and, among the literate classes, expressed in diaries and letters to mothers, sisters and other female relatives, as women advised, prepared and consoled each other. During the delivery, the midwife and gossips in attendance would take care of all aspects of the birth and lying-in.[18] As part of a process already in place by the eighteenth century and accelerating into the nineteenth, this was changing, as birth increasingly became the concern of doctors, and women were eased out and excluded from the birthing room as mid-wives and helpers.[19] Although doctors could potentially become women's allies in childbirth, responding to requests for pain relief and intervention to shorten labour, they also stripped away a source of emotional and practical support, banishing female helpers as outmoded and potentially dangerous. The council of women, which had tended to rely on 'a just reliance on the powers of nature',[20] was replaced by the council of men, men-midwives and obstetric doctors, through their consultations and through advice literature. Advice books, intended to fill what came to be seen as a lacuna of information on pregnancy, confinement and care of the new-born, stressed that motherhood was women's noblest function – and that women would be failing in their duties if they did not prepare themselves as best as they could for delivery. Thomas Bull, author of one of the most popular advice books of the nineteenth century, wrote much in his book that was plain common sense and useful advice but prefaced it with the injunction:

> A woman may consider herself a mother, not only from the birth of her child, but even from the moment of conception. From that important epoch her duties commence – duties amongst the most sacred and dignified which humanity is called upon to perform . . . Should, she, however, be careless and negligent upon this head, and fail in attention to the measures which her new condition demands . . . her child will inevitably be variously and injuriously affected, these causes operating through her system upon that of the child.[21]

Advice book literature thus applied an emotive language to the responsibilities of mothers and the risks they took in neglecting their natural duty, spelling out in sometimes chilling detail the impact their carelessness or inattention could have. Martha Mears, writing at the commencement of the century, was largely supportive and optimistic about the process of childbirth, but also related how fear in pregnancy chilled the vital parts and 'clogs the fine machinery of the whole frame'. Fear could increase the difficulty of childbirth, cause debility and disease and result in 'a puny, or distorted infant', 'the victim of its mother's terrors'.[22] Doctors stepped up this emphasis on the negative impact of the wrong kind of emotions among expectant mothers, which could have a deleterious effect on the newborn child. While denouncing the outmoded idea of maternal imagination, Bull argued that women who were unable to maintain calm throughout pregnancy risked injury to their child. He cited the example of a woman greatly depressed during pregnancy due to the 'worry' of her husband about supporting another child. 'In consequence of this mental harass[ment] and disturbance, she was confined shortly after the completion of the seventh month. The child was puny and fretful, and continues so.' At eight months it was a 'wasted miserable-looking object', who rarely smiled.[23] The proliferation of literature of this kind indicates – and it appears to have impacted strongly, with Bull, for example, going through numerous editions within a few years – that it was making its mark on well-to-do households, the kind most likely to employ a male attendant well versed in the literature on hazards associated with childbirth.

Debilitating and dangerous conditions linked to childbirth had been described and elaborated long before 1800. After this date, however, the sheer intensity and extensiveness of writing on the subject was intended to impress upon childbearing women and their attendants the gravity of the event they were experiencing. At the same time a powerful link was being forged between obstetrics, gynaecology and psychiatry, as women's nervous and mental disorders were associated increasingly with reproduction. Thomas Denman, writing at the turn of the nineteenth century, was the first British midwifery practitioner to dedicate serious attention to the topic of mental disorder in childbirth, which he named 'mania lactea'. Yet he did not see mental disorder as something that could be closely defined or as an anticipated sequel to childbirth; arguing that pregnancy, labour and childbed represented 'an altered, but not a morbid state'.[24] His successors, however, would raise expectations of mental collapse in association with delivery. Puerperal insanity found a special place among the burgeoning spectrum of female disorders, linking as it did the bodily upsets of childbirth to mental disturbance. As Robert Gooch, who coined the term 'puerperal insanity' in 1820, and was subsequently acclaimed as the chief authority on the subject, explained:

> During that long process, or rather succession of processes, in which the sexual organs of the human female are employed in forming; lodging; expelling, and lastly feeding the offspring, there is no time at which the mind may not become disordered; but there are two periods at which this is chiefly liable to occur, the one soon after delivery when the body is sustaining the effects of labour, the other several months afterwards, when the body is sustaining the effects of nursing.[25]

Puerperal insanity became a hot topic for obstetric practitioners and alienists following the publication of Gooch's treatise on the subject in 1820.[26] For some doctors, puerperal insanity was linked to morbid disease or a predisposition to mental illness, or related to hereditary influences. For others, it came to be seen as a straightforward risk of becoming a mother, the result of the very act of giving birth, a risk which commenced with conception and ended only when the child was weaned. As George Man Burrows, who linked the fields of obstetrics and psychiatry in his own career trajectory, having first practised midwifery before specializing in mental alienation, wrote in 1828:

> Gestation itself is a source of excitation in most women, and sometimes provokes mental derangement, and more especially in those with an [*sic*] hereditary predisposition. The accession of mental disturbance may be coincident with conception, and cease on quickening; or it may come on at any time during pregnancy, continue through it, and terminate with delivery, or persevere through all the circumstances consequent on parturition. Some are insane in every pregnancy or lying-in, others only occasionally.[27]

As Anne Digby has argued, such descriptions would take on new meanings in the nineteenth century.[28] The social context and ever closer association of obstetrics and psychiatry, combined with redefined notions of women's proper role and functions, would bring a new perspective to the link between childbirth and mental collapse, associating it with deeper fears about women's general vulnerability.

The landscape and language of emotion: Framing puerperal insanity

Puerperal insanity, however, would never be described exclusively as a 'medical' condition; its onset, as will be explored in the following pages, was linked to a wide variety of events and circumstances, with childbirth sometimes featuring as a mere adjunct or trigger at the end of a long line of predisposing factors. It was, however, defined increasingly in medical terms, which partially replaced (although far from conclusively), the more day-to-day, non-medical language of earlier centuries, when mental disturbance in childbirth was attributed to anxiety, fright or poor management of the delivery which resulted in a 'dejection of spirits' or 'passions of the mind'.[29] The emergence of puerperal insanity in the nineteenth century appears to have been marked, too, by a loosening of a link with such passions, which were replaced with a new set of pathologized, medicalized emotions.[30] Again this was a very partial loosening, and, though no longer described as 'passions', per se, passionate behaviour was a major aspect of the typology of puerperal insanity in terms of causes and symptoms.

As increasing numbers of patients were confined to asylums in the nineteenth century, numbers were produced to highlight the frequency with which puerperal insanity occurred. This in turn bolstered the idea of a collective risk being attached to childbirth; puerperal insanity was observable in groups of patients showing similar symptoms, rather than individual instances of mental breakdown. By the mid-nineteenth century, John Charles Bucknill and Daniel Hack Tuke suggested that the proportion of patients described as 'puerperal' varied from around one-eighth to one-twentieth of female admissions.[31] Between 1843 and 1848, Bethlem hospital admitted 899 patients, of whom 111 (or 12 per cent) were diagnosed as suffering from puerperal disorders.[32] Not only were these cases countable, but they were also relatively easy to identify with their link, even if this was in some cases tenuous, to childbirth, as puerperal insanity emerged as 'one of the few clearly recognized entities in 19th century psychiatry'.[33] Bucknill and Tuke's influential textbook acted as a guide to many asylum superintendents, with its emphasis on the prevalence of puerperal insanity, and its usefulness as a classificatory category, with sub-divisions into insanity of pregnancy, puerperal insanity and insanity of lactation.[34]

Just as women found themselves exposed to shifting medical paradigms, with sub-texts of anxiety and fear and an increasing expectation that childbirth might result in madness – so not just something to be observed and acted upon, but something to be watched for and guarded against – male practitioners, too, were exposed to a new set of dialogues and experiences. Many doctors in the nineteenth century worked increasingly with female patients, particularly in midwifery.[35] Puerperal insanity exposed different groups of practitioners, obstetric doctors and alienists, to female patients in the home, the maternity hospital, the workhouse lying-in ward and the asylum. It has been claimed that the Victorian period saw the takeover of childbirth practice by doctors, with professional manipulation, exclusion of the midwife, a stressing of knowledge over experience and cynical hard-nosed survival to make a medical living.[36] It could be argued that it also exposed doctors to direct experiences of childbirth and to the bodies and minds of women. These exposures could impact dramatically on the doctors involved, which their writing on cases of puerperal insanity makes explicit, yet the potential emotional impact on male practitioners has tended to be overlooked by historians of medicine. Examining the descriptions of puerperal insanity, particularly case histories, provides evidence of the ways in which motherhood was configured as a rich, though often negative, emotional experience. It also gives an indication of the responses of frequently young and inexperienced doctors, when confronted with the mental breakdown of their patients following childbirth. Case histories testify not only to the complex mental and emotional experience of mothers, but also express doctors' own concerns and distress.

Though troubled about the impact that the mental breakdown of patients following childbirth might have on their reputation and practices, it was not without sincerity that doctors described how the condition marred a scene of potential happiness to the mother and her family. 'Confinement' took on a poignant meaning when connected with puerperal insanity. The happy event of giving birth was tainted by the illness of the mother and the disruption of family life. Childbearing women were depicted as being vulnerable to a range of mental and physical challenges and disorders and childbirth was expected to be a dangerous passage, yet puerperal insanity was represented as a particularly cruel twist of fate and in many women unexpected and unaccountable given their previous medical, moral and family history. Fleetwood Churchill described the condition as 'a very distressing malady in itself, but doubly so from occurring at a moment ordinarily so joyful'.[37] George Man Burrows portrayed the impact of puerperal insanity in terms of the reversal of the expectations of a happy delivery:

> Fondly anticipating the joy, perhaps of a first-born, a beloved wife patiently submits of all the inconveniences and restraints of pregnancy, however irksome, and the pains and dangers of labour, however great. The affectionate husband and relatives await with deep and anxious expectation the event; and at length, when the joyful period arrives, and the happiness of all is completed by a safe delivery, – how dreadfully is the scene reversed, when the happy mother suddenly displays symptoms of delirium![38]

Though often frustrating to deal with, not least because of the violence of their symptoms, patients were dealt with sympathetically by their medical attendants and were coaxed rather than bullied back to health. Heroic remedies and drugs were generally derided, and treatment regimes relied on diet, the encouragement of rest and sleep, and moral management, urging regular patterns of deportment and activity.[39]

The application of classificatory labels and the appearance of countable numbers were paralleled by a stepping up in the production of medical explanations for puerperal insanity. Yet throughout the century it was defined and explained in a variety of ways, many non-medical, allowing for a rich discourse and evocative language to emerge outlining the perils of motherhood. It would be difficult to argue that puerperal insanity ever became a closely defined disorder – it was notable for its inclusiveness of melancholic and manic forms and mild and severe cases, and for the range of women affected, young and old mothers, rich and poor, married and unmarried, women undergoing their first confinement and the mothers of many children, women undergoing a normal and trouble free birth and those suffering terrible labours, painful, lengthy and instrumental deliveries – and for the variety of predisposing factors and prompts which led women to fall prey to it.[40]

Puerperal insanity was also identifiable by a range of behaviour that had much in common with the very affections and passions, which, it has been argued, were being replaced by an overarching category of emotions.[41] The women were violent and destructive, giving free rein to a range of unfeminine passions and actions, tearing their clothes, attacking family members, servants and their medical attendants, speaking of strange events, swearing and cursing, making sexual advances, overturning furniture and smashing windows. The sheer physicality of the experience of puerperal insanity is alluded to frequently in case notes, the women's power and strength forming a distinct contrast to maternal images emphasizing succour and gentleness. Jane Anderson was admitted to the Royal Edinburgh Asylum in 1848 after murdering her seven-week-old child. After her confinement she became seriously disturbed:

> She started suddenly out of sleep in a state of terror, fancying that she was about to be torn to pieces by wild beasts. Sometime she fancied that she was about to be carried away and put to some violent death . . . She was allowed to remain in her own house carefully watched, until this morning, where she murdered her Infant by cutting its throat with a Razor.[42]

The language of religion provided women with a vivid means of expression, and their delusions often referred to the work of the devil, stealing their souls and provoking them to carry out violent acts. Agnes

Watson was admitted to the Royal Edinburgh Asylum in May 1865 after several months of nursing an illegitimate child who had subsequently died. She was very violent and threatening to others, and refused food 'because "a dead person requires none"'. She imagined that she was being 'devoured by a foul spirit', and that the devil attempted to take her off her bed and throw her out of the window: 'the devil tells her constantly he had done for her now'.[43]

Case notes of asylums and private practice are rich sources for exploring the language used to describe puerperal insanity and explain its occurrence. They could be candid about the problems of treating the condition and the difficulties, given the multiplicity of causes, of preventing its occurrence and reoccurrence. The case notes framed women's flight into insanity as rich and varied emotional experiences. Far from being merely 'straitjacketed' by cultural and medical dialogues making reference to limited female roles, feelings and possibilities defined by bodily and social restrictions (primarily their duties as mothers at the centre of the household), women were described as experiencing emotions that could reflect disappointment with their roles, unhappiness, frustrations and also ambitions and hopes. Childbirth was represented as a paradox. Bearing children was an almost sacred duty, women's most important function, but one that exposed them to a multiplicity of bodily risks, and which also raised their levels of sensitivity and sensibility, and predisposed them to mental breakdown. True, their heightened emotional states were seen as ultimately destructive and increasingly too as a medical event, which challenged 'normal' behaviour. Nonetheless, these women were also seen as having fertile imaginations and feelings. Doctors described mothers as being subject to a range of emotional states, many of which were held responsible for precipitating their madness, and which were embedded in life circumstances and family relationships.

Doctors confronted with cases of puerperal insanity typically put together complex explanations for the women's loss of reason based on a collection of factors rather than a specific medical event. Childbirth was the trigger for puerperal insanity, but the build up to this was often set within a much longer time frame. The obstetric physician James Reid concluded that 'vast changes in the uterine organs during pregnancy, and more rapidly immediately after parturition, there is superadded an acknowledged state of great nervous excitability . . . and above all, the influence of *moral* causes to so great an amount'. Reid added that, given the potential for physical and moral factors to combine, he was surprised that puerperal insanity was not more frequently observed.[44] And indeed explanations for women's susceptibility to puerperal insanity included, but also went well beyond, the biological, stressing a range of factors from the very obvious ones of hardship, overwork, weariness, poverty and poor physical health to the less tangible – ill-treatment or unhappiness, domestic crises, a lack of preparedness or anxieties about motherhood, over-sensibility and disappointment. Many of the preoccupations described by doctors crossed class boundaries, notably the anxieties expressed by women about becoming mothers or disappointment about their life situation. Even well-to-do women became obsessed with notions of impending financial disaster and impoverishment, and fears that they in some way were responsible for the ruin of their families. Gooch described the case of one woman who was admitted to a private madhouse in London suffering from puerperal mania. She was suicidal and was convinced that she had committed crimes that had brought ruin and disgrace on herself and her family, which had occasioned the death of her husband and children.[45] Husbands were frequently accused in case notes of endangering the mental health of their spouses, described as womanisers, bullies or drunkards, and women of all social classes were said to be disappointed in their husband's failure to provide properly for them. Robert Boyd, for many years physician to the Somerset County Lunatic Asylum, believed that puerperal insanity was distinguished as much by the women's hatred of their husband as their new infants.[46]

Extensive use was made of the term 'irritability' to describe both the physical condition and mental status of women poised to succumb to puerperal insanity. It was a curiously flexible term, straying well beyond medical applications, used to refer to a more general irritation with family life and annoying circumstances. Gooch described, for instance, how

> Nervous irritation is very common after delivery, more especially among fashionable ladies, and this may exist in any degree between mere peevishness and downright madness. Some women, though naturally amiable and good tempered, are so irritable after delivery that their husbands cannot enter their bedrooms without getting a certain lecture; others are thoroughly mad.[47]

He concluded that the 'peculiar state of the sexual system which occurs after delivery' diffused an 'unusual excitement throughout the nervous system' producing the disease in predisposed women. The female constitution was so weakened by labour, and the nervous system so overwrought, disorders of the mind easily supervened.[48] The midwifery practitioner John Burns concluded that 'All women, in the puerperal state, are more irritable, and more easily affected, both in body and mind, than at other times, and some even become delirious'.[49] Marshall Hall deemed puerperal mania as 'a mixed case', caused by 'all circumstances following parturition combined', but chiefly intestinal irritation, loss of blood and exhaustion.[50] Women, with their systems disturbed by childbirth and frequently in a state of sleeplessness, were particularly vulnerable to irritations or minor shocks; annoying nurses or relatives disturbing the peace of the lying-in room, ringing doorbells, the appearance of strangers, fires in the neighbourhood were all implicated in cases of puerperal mania. One woman, still weak after her first, very lengthy labour, was reported as having succumbed to madness, followed 13 days later by her death, after the receipt of a letter from her husband declaring his bankruptcy.[51] The obstetrician James Simpson described a more remarkable incidence where a bout of insanity was triggered by the 'cruel gift' of a box of mice, given by an 'evil-minded relative' to a lady convalescing after childbirth who had a 'constitutional horror' of the creatures.[52]

Accounts of puerperal insanity turned frequently on the heightened emotional state of women, their increased susceptibility not only in terms of their bodily states, but also the actions of their imagination, dreams, perceptions and even their intellect. John Tricker Conquest spoke of the higher incidence of puerperal insanity in 'females of extreme sensibility, whose mental or physical powers dispose them to be inordinately influenced by causes which would scarcely affect other women, or even themselves, but for the susceptibility to disease, and the peculiarity of condition consequent to delivery'.[53] Robert Gooch explained how puerperal insanity was more likely to occur in patients 'of susceptible dispositions, nervous, remarkable for an unusual degree of that peculiarity of nerve and of mind, which distinguishes the female from the male constitution'.[54] Over-cultivation and 'bookishness' could place women in higher-risk categories, according to Gooch, who cited the case of a woman suffering from puerperal mania, who was 'clever, susceptible and given to books' and another who had amused herself during labour by reciting poetry'.[55]

Asylum casebooks, almost without exception, described patients as living in extreme poverty and having very poor health. The women were depicted as having unpleasant, exhausting and disordered lives, and giving birth added an even greater burden. Yet, even considering the dire situations of these women, the language employed to describe them is striking, with the casebooks littered with words such as 'brooding', 'desponding', 'disappointed', 'harrassed' and 'desolate', many referring to unpleasant life circumstances rather than any medical event. Stress or anxiety were said to play a large part in the women's derangement. Thus not only well-to-do sensitive women were accorded rich emotional lives, but also poor, hard-working mothers. Harriet Ashmore had already 'been peculiar' for two months, 'desponding and threatening suicide', before giving birth to her sixth child. Immediately after being confined she became much worse, even though she had a 'normal and easy' delivery:

> For a considerable period before her accouchement, she was harassed by debt, and had to contend with great poverty, her means were very insufficient for the maintenance of herself and family so that she had to deny herself many of the necessaries of life, and was therefore weak and badly nourished when her confinement came on. Her excitement has partaken of a suspicious character, and her delusions have been chiefly about the persecutions to which she is subjected and the misfortunes that have befallen her family.[56]

Fear loomed large in descriptions of the onset of puerperal insanity, with dread of the birth itself intermingling with anxiety concerning family resources. Emma Walker's case apparently greatly perturbed Henry Parsey, the Superintendent of Warwick County Lunatic Asylum, who was used to seeing women admitted in an appalling physical state. Walker was not even of the poorest class of patients, but her mental anxiety was extreme. A 33-year-old dressmaker, she was admitted to the asylum in November 1864 after having been insane for about three weeks. She was confined with her fifth child about six months earlier, and 'previous to that had been much harassed and distressed by her husbands [*sic*] embarrassed circumstances and threatened bankruptcy'. She did not make a good recovery after delivery:

> She was still kept in constant anxiety and fear, and took to her old employment, dressmaking, with the view of eking out their means. She nursed her child up 'till a fortnight ago when the secretion of milk suddenly ceased. A week before that a change had been noticed in her manner and disposition. She had become very irritable and capricious, had exhibited dislike to her children and had partaken of very little food.[57]

Before admission she became increasingly excited, restless and incoherent and very destructive, had little nourishment and no 'natural' sleep. She was delusional and attempted suicide. The case notes referred to the death of one of her children during a previous marriage, which still preyed on her mind. Emma Walker had been forcibly restrained in a strait-waistcoat before being brought to Warwick and was found to be covered in sores. The sores rankled with Parsey, an advocate of moral management and non-restraint, and he was also aware that the woman's mental distress was due to much more than straightforward medical causes. Indeed, her delivery is only referred to in passing as part of a troubling account of anxiety and stress linked to maternity and family life.

Many women were depicted as being overly anxious about their ability to be good mothers, expressing worries about breast-feeding their infants, that were no doubt stimulated by the fact that many newborn children failed to thrive. John Batty Tuke, based on his experience at the Royal Edinburgh Asylum, claimed that first-time mothers were most likely to develop insanity of pregnancy, which he attributed to 'the moral exciting causes, anxiety and dread of the coming event, which exist to a greater degree in the inexperienced woman'.[58] Such fears were framed by doctors as part and parcel of the emotional ambiguities surrounding motherhood, which could be marked by distress and worry as much as joy. Maternity was presented as a severe physical and mental challenge. Though ideally a blessing and a source of fulfilment, it could confront women with hard work and worry about their new responsibilities as well as making them cross and difficult to live with. William Ellis, the Medical Superintendent at Hanwell and formerly of the West Riding Asylum at Wakefield, concluded 'there is no doubt, but that the various circumstances of hope and fear in which females are necessarily placed at such times, render them more sensitive than usual to the operations of a variety of moral impressions'.[59]

Emotional communities: The household and puerperal insanity

Puerperal insanity had an intimate relationship with the household.[60] The disorder was unsurpassed in its ability to challenge the hegemony of domestic ideology, as the household was disrupted, turned upside down, social mores and notions of domestic order challenged in a noisy and forceful way, as women became alarming spectres in their own homes. The ultimate aim of a cure was to place women back into their prescribed roles as housekeepers and mothers. Yet the home, too, was depicted as the root of the disorder, the burdens of motherhood and household responsibilities as the actual causes of distress. There was also considerable debate about the role of the home as a place of treatment. Rather more than was the case for other forms of mental illness, puerperal insanity was a disorder that was treated in different settings, and there was a lively discussion throughout the nineteenth century about whether home or asylum was the most appropriate place.

Working loosely with Barbara Rosenwein's concept of 'emotional communities' and 'systems of feelings',[61] the final part of this essay will focus briefly on the experiences of patients suffering from puerperal insanity within the household and the interactions of patient, family and doctors, and how the emotional condition of puerperal insanity was treated. According to Rosenwein, looking at emotional communities enables the uncovering of systems of feeling: 'what these communities (and the individuals within them) define and assess as valuable or harmful to them; the evaluations that they make about others' emotions; the nature of the affective bonds between people that they recognize; and the modes of emotional expression that they expect, encourage, tolerate and deplore'.[62] Rosenwein suggests an approach to inner life in the past that is also capable of accommodating historians' concerns with time and place – without necessitating the acceptance of modernizing narratives (popularized by Norbert Elias and his followers), in which 'the history of the West is the history of increasing emotional restraint'.[63] Her model also, significantly, moves the literature on 'emotional restraint' beyond the terrain mapped by theorists such as Goffman and Foucault in their descriptions of institutional spaces and power relationships within them.[64] The ultimate place of restraint in terms of confinement (and, according to Foucault, emotional straitjacketing), the asylum, holds great potential as an emotional community;[65] the other, in the case of puerperal insanity, is the patient's own home.

Within the household, predominantly those described in case notes as upper- or middle-class, doctors can be seen as not only recording and responding to emotional displays, but also as actively contributing to a dialogue. They describe themselves as directors and lead actors in scenes of great emotional intensity. Their role was to direct the other actors, the patient and her family, but also to take the lead in curing the woman and putting the household to rights. As puerperal insanity was being defined in the early part of the nineteenth century as a new disorder, and was treated in different contexts by different sets of practitioners all claiming expertise, their discussions are suggestive of considerable eagerness to frame and understand the condition and their place in treating it. So in this way they could be regarded as scriptwriters as well. The descriptions contained in case notes embody great dynamism as well as a heightened sense of observation.

Over the course of the nineteenth century, the asylum became the ultimate place of observation and treatment of puerperal insanity, but a great deal of ambiguity was expressed – by alienists as well as obstetric practitioners – about where the patient would do best. Many doctors argued that so long as the immediate damaging influences of infant and family were separated from the woman, she was most likely to thrive in a domestic environment, cared for by nurses supervised by the medical attendant. By the 1840s, the renowned alienist John Conolly was advocating the asylum as the only appropriate place to treat cases of puerperal insanity and to develop knowledge of the condition, but this constituted a marked shift from the views he expressed earlier in his career when he argued that the asylum would be detrimental to such cases: 'To separate her from her infant and her family, and to place her among strangers, is to debar her from almost every hope of being soothed and calmed.'[66] Conolly regarded puerperal insanity as a particularly difficult disorder for a medical practitioner to deal with as the household dissolved into chaos, and he advised emotional restraint in responding to a highly charged situation:

> Few cases can be supposed to occasion more distress in a family than the unexpected appearance of insanity in a young woman, just when she, her husband, her relatives, and her friends, are full of natural joy on account of her having safely become a mother. A more than ordinary weight of anxiety is consequently thrown upon the practitioner, who is earnestly besought to answer interrogatories for which his practical experience has scarcely prepared him. He is questioned in whispers at the door of the bedroom, intercepted on the stairs, and cross-examined by an anxious assembly in the drawingroom ... he is hoping that the anxious part of the case is at an end, and he has to meet new difficulties, against which he finds little aid in a house now thrown into confusion and with monthly nurses seeming, in these new circumstances, without resources, and full of alarm. It is well if these perplexities do not drive him to hasty and injudicious proceedings.[67]

Robert Gooch's case notes, with their detailed and empathetic observation of the interactions between family members and patients, open up stories of the impact of puerperal insanity in the middle-class household, emphasizing the chaos, the upset, and the emotional changes taking place in his patients.[68] Though Gooch held true to values of order and domesticity, he also expressed sympathy, even admiration, for the heightened emotional experiences of his patients; he was at pains to emphasize that the disorder represented an aberration and normally his patients were upstanding women, fully committed to their household duties. Gooch also used case histories as a means of building his knowledge, with his patients' stories of their physical discomfort, weariness and mental distress enabling him to place diverse symptoms into this new and distinct disease category. The women he described attempted to keep a grip on the household and their duties, but finally, inevitably, domestic order was replaced by misrule:

> One of my patients, almost from the day of her delivery, was observed to have restless nights, a quick pulse, and an irritable temper, compared with her natural one. She scolded the nurse about the merest trifles: one minute sent for the child to suckle, and the next ordered it angrily to be taken away. She would superintend her housekeeping, though she was entreated not to do it, and sent for the cook up into her chamber several times a day to enquire into the consumption of the family, and to give directions about its regulation. She talked almost incessantly, with disproportionate earnestness; and complained that her husband was not attentive to her: at length she accused him of incredible things, and soon after became so violent as to require confinement.[69]

Gooch and other commentators recognized sleeplessness as an early symptom of puerperal insanity, but so too was a deviation in the patient's language or behaviour, especially if she became quarrelsome, critical, peevish, fretful, sullen or discontented.[70] Such infringements of codes of conduct and duty were reckoned to be as telling as the more violent outbursts that could typify puerperal insanity. Gooch responded to his patients' cases with the decorum that was to become a reputed hallmark of his practice; in times of crisis he opted to remain in the household, often overnight, so that he could closely observe his patients. Even when reporting cases where the household was reduced to turmoil, Gooch pointed to the woman's naturally good demeanour and pleasant character. The physician's role should centre, he claimed, on efforts to treat and cure the woman, but he should also act as a friend and diplomat.

The links between disorder, domestic chaos and inappropriate behaviour would be remarked upon and developed in many accounts of puerperal insanity. The account of Robert West, a rural general practitioner, illuminates the impact of a household in chaos and also the impact of a wide range of non-medical causes and upsetting circumstances on such cases. West recognized the extent of Mrs V's disturbance, a farmer's wife he had attended in labour, when five days after the delivery she 'actually got out of bed, proceeded down stairs and into the yard, with nothing on, but her night-clothes, and without shoes; and had helped herself to a copious draught of cold water at the pump!'[71] West, who described the case in a letter to his midwifery lecturer, Dr Ryan, largely attributed her attack of puerperal mania to the circumstances of the lying-in:

> Her mother, who was to nurse her, did not arrive until eleven o'clock in the evening, and then found no one in the house with the patient but a little girl. The fires up stairs and down were all out. For the four following days, the mother had all the work of the house to do, cows to milk, calves to feed, and I don't know what, besides attending to her nurselings; Mrs. V. all the time lamenting that her mother had much to do. A servant had been hired, but had failed to come, which was another source of mental disquietude.[72]

On the third day Mrs V sat up for an hour, and on the fourth for four hours, 'being all the time surrounded by gossiping friends, and being driven to bed at last by the pain in her head ... It was said that she drank a quantity of rum on one of these days'.[73] The woman died shortly afterwards. West was 'sadly troubled' about the case but blamed the woman's madness and death on the lack of care she had received and the disturbance to the household.

Doctors treating puerperal insanity were managing an emotional space or situation, and the clearest expression of this is found in home settings, where they mediated directly on a day-to-day basis between the patient, her illness and her family. Disorder resulted from the affliction of the mother with puerperal mania, but the illness could also have its root in disrupted households, or labours disrupted by friends who thought it a time of merriment, in Gooch's words 'footing it about the house, which resembled a rabbit warren'.[74] Though the need for sympathy and patience was stressed, so too was the need to assert authority, on the part of nurses and particularly the medical attendant, in order to cure the patient and put the household to rights. As Dixon has argued, this was vital to the physician who 'must be able to command and govern badly behaved patients, who, like his passions, had the potential to "ruffle his temper, and consequently cloud his judgment, and make him forget propriety and decency of behaviour"'.[75] In 1848 Arthur M'Clintock and Samuel Hardy, both formerly of the Dublin Lying-in Hospital, passed on their techniques of patient control, offering the following advice if the woman proved refractory, when 'the contest will be whether she or the doctor is to have the ascendancy':

> Under these circumstances it is requisite for the physician to exercise much tact and resolution in his language and conduct towards the patient. If he does not succeed in enforcing his directions by mild expostulation, he must shew himself to be determined, and, without harshness, insist on his orders being obeyed; for if, through vacillation, or want of resolution, he now fails to establish his authority, he will lose all control or restraint over the patient; whereas, on the other hand, if he carry his point, and bring her into compliance, it will have a lasting effect, and she will probably stand in awe of him during the remainder of her illness.[76]

The role of family members was typically reduced to inspecting the arrangements concerning treatment. Husbands, often the butt of hatred on the part of female sufferers, were decisively pushed out of the therapeutic regime. Even in wealthy households, wives were removed from their husbands' presence and authority. The authority of the doctor in attendance, who replaced the husband, and insisted on removing friends, relatives, familiar nurses and servants, reigned supreme, and gained the time and space needed to effect a cure: 'It is necessary to gain, by firmness, a mastery over such patients. They recover more quickly when not allowed to see their husbands, infants or immediate relatives, and they are most easily managed by nurses whom they have not previously known.'[77]

Puerperal insanity was a severe form of mental illness, but it is challenging when reading case notes and other forms of documentation to establish a clear borderline between intense expressions of emotion and mental disorder. Bucknill and Tuke, with their keen interest in moral or 'stress models' of insanity, suggested that mental instability and stability lay along a continuum, implying that, with a combination of stress and predisposition, anyone could fall victim to insanity.[78] As women were recognized as being exposed to a range of challenging circumstances and strains around childbirth, which added to the physical stress on their bodies and nervous systems, puerperal insanity came to be seen as a disorder that was anticipated or expected to occur. The condition could establish itself over a long period, particularly cases of melancholia, and its onset was often marked by feelings of sleeplessness and exhaustion, loss of control, extreme anxiety, fear and depression, and in some cases by the anticipation on the part of the woman of an impending decline into insanity. As Gooch explained, the approach of the disease excited little apprehension and the symptoms were very diverse; the women's pulse could be quick, their nights restless and temper short, nothing remarkable. Gooch referred to what followed as 'indescribable hurry' and 'peculiarity of manner', the women's language becoming wild and incoherent as they reached a state of mania. When the disease took the melancholic form it began some months after delivery; the patients typically declined in health while nursing, suffering from failing memory, confusion and a depression of spirits. They found it difficult to concentrate and were bewildered, anxious and dissatisfied about themselves.[79] Many of the feelings and symptoms – anxiety, tiredness and distracted behaviour – signifying the onset of the condition of puerperal insanity were also part and parcel of the experience of childbirth, and it is not clear how much emotional intensity was required to result in the diagnosis of insanity. Though some patients were declared frantic and violent in their emotions, others were merely subdued and miserable. At such a fraught time as childbirth, puerperal insanity occupied a grey zone between mental illness and highly emotive behaviour; there was great opportunity for slippage from one to the other.

Explanations for the onset of the condition, and for some women's particular susceptibility to it, were set, and continued to be so for much of the nineteenth century, within contexts of hardship, dissatisfaction, misery, pressing worries and doubts.[80] Although the language of medicine intervened massively in framing puerperal insanity, those treating it clung to a richer terminology and explanatory framework in explaining its prevalence. While the case notes of private practice, dealing with cases within household settings, more often than not refer to wealthy patients, the poor were not excluded from the impact of upsetting and alarming emotions associated with their roles as housewives and mothers. Doctors saw themselves as being up against something much larger than women's biological vulnerability, witnessed women's energetic and full emotional lives, and in particular recognized that motherhood could be disruptive, overdemanding, overwhelming and disappointing. We need to be cautious about oversimplifying models of social roles and of medicine in a period marked more by openness to a wide range of ideas and explanatory frameworks for the mental breakdown of new mothers rather than by the imposition of new and rigid ideologies.

Notes

1. Dr Fay Bound Alberti offered very useful commentary on an earlier draft of this essay, and I would like to thank the participants at the workshop in London on which this volume is based for useful feedback on my paper, and to Professor Margot Finn for conversations on emotions and for organizing a seminar series on the history of emotions at Warwick University. Much of the research for this paper was carried out while I was holding a Wellcome Trust University Award, and I would like to thank the Wellcome Trust for their support of this work.
2. Judith Schneid Lewis, *In the Family Way: Childbearing in the British Aristocracy, 1760–1860* (New Brunswick, NJ: Rutgers, 1986), p. 62.
3. William Buchan, *Domestic Medicine; or Treatise on the Prevention and Cure of Diseases*, by Regimen and *Simple Medicine* (Newcastle: K. Anderson, 1812 [1772]), p. 28.
4. Christian Augustus Struve, *A Familiar Treatise on the Physical Education of Children*, translated from German and with three lectures by Anthony Florian Madinger Willich, MD. (London: Murray & Highley, 1801) p. 121. Cited in Lewis, *In the Family Way*, p. 63.
5. Sally Shuttleworth, 'Demonic Mothers: Ideologies of Bourgeois Motherhood in the Mid-Victorian Era', in Linda M. Shires (ed.), *Rewriting the Victorians: Theory, History and the Politics of Gender* (London and New York: Routledge, 1992), pp. 31–51, 31.
6. Eardley Holland, 'The Princess Charlotte of Wales: A Triple Obstetric Tragedy', *Journal of Obstetrics & Gynaecology of the British Empire* (1951), 58, pp. 905–19. The triple obstetric tragedy refers to the death of Princess Charlotte, her baby and the subsequent suicide of her obstetrician attendant Richard Croft. See also Lewis, *In the Family Way*, pp. 182–7.
7. See Hilary Marland, *Dangerous Motherhood: Insanity and Childbirth in Victorian Britain* (Houndmills: PalgraveMacmillan, 2004), esp. chapter 2.
8. George Man Burrows, *Commentaries on the Causes, Forms, Symptoms, and Treatment, Moral and Medical, of Insanity* (London: Thomas and George Underwood, 1828), p. 368.
9. James Reid, 'On the Causes, Symptoms, and Treatment of Puerperal Insanity', *Journal of Psychological Medicine and Mental Pathology* (1848), 1, pp. 128–51, 284–94, 134–5.
10. Cited in Anne Lawrence, *Women in England 1500–1700: A Social History* (London: Phoenix, 1996), p. 80.
11. See e.g. Mark Jackson, *New-Born Child Murder* (Manchester and New York: Manchester University Press, 1996), esp. pp. 120–3.
12. Sir John Bickerton Williams, *Memoirs of the Life and Character of Mrs Savage* (London, 1821) p. 286. Quoted in Patricia Crawford, 'The Construction and Experience of Maternity in Seventeenth-Century England', in Valerie Fildes (ed.), *Women as Mothers in Pre-Industrial England* (London and New York: Routledge, 1990), pp. 3–38, 22. See also Linda Pollock, 'Embarking on a Rough Passage: The Experience of Pregnancy in Early-Modern Society', in Fildes (ed.), *Women as Mothers*, pp. 39–67.
13. County Record Office, Carlisle, D/KEN. 3/56/1 (c.1801), conduct letter written by E. Kennedy. Quoted in Amanda Vickery, *The Gentleman's Daughter:*

Women's Lives in Georgian England (New Haven and London: Yale University Press, 1998), p. 98. See also Lewis, *In the Family Way*, pp. 74–5 for the penning of wills and farewell letters from noble women to their husbands, and for fears of childbirth expressed by American women, Judith Walzer Leavitt and Whitney Walton, '"Down to Death's Door": Women's Perceptions of Childbirth in America', in Judith Walzer Leavitt (ed.), *Women and Health in America* (Madison, WI: University of Wisconsin Press, 1984), pp. 155–65.

14. Martha Mears, *The Midwife's Candid Advice to the Fair Sex; or the Pupil of Nature* (London: Crosby and Co. and R. Faulder, c.1805), p. 28.
15. John Woodward, *Select Cases, and Consultations in Physick. By the Late Eminent John Woodward* (London: Davis & Reymers, 1757), pp. 259–65. The full case, from which this summary is drawn, is published in Richard Hunter and Ida Macapline (eds), *Three Hundred Years of Psychiatry 1535–1860* (London: Oxford University Press, 1963), pp. 338–41.
16. Ibid. Cited in Hunter and Macalpine (eds), *Three Hundred Years of Psychiatry*, p. 340.
17. Robert Lee, *Lectures on the Theory and Practice of Midwifery, Delivered in the Theatre of St. George's Hospital* (London: Longman, Brown, Green, and Longmans, 1844). This was also reported in the London Medical Gazette, 1842–43, p. 1.
18. For networking and cultural links between mothers and midwives, see Doreen Evenden, *The Midwives of Seventeenth-Century London* (Cambridge: Cambridge University Press, 2000), and for the organization of labour and the lying-in by midwives, Adrian Wilson, 'The Ceremony of Childbirth and its Interpretation', in Fildes (ed.), *Women as Mothers*, pp. 68–107.
19. See e.g. Jean Donnison, *Midwives and Medical Men: A History of Inter-Professional Rivalries and Women's Rights*, 2nd ed. (New Barnett: Historical Publications, 1988 [1977]).
20. Mears, *The Midwife's Candid Advice*, p. 2.
21. Thomas Bull, *Hints to Mothers, for the Management of Health during the Period of Pregnancy and Lying-In Room; with an Exposure of Popular Errors in Connection with Those Subjects*, 16th ed. (London: Longmans, Green, and Co., 1865 [1837]), pp. 3–5.
22. Mears, *The Midwife's Candid Advice*, pp. 27–8.
23. Bull, *Hints to Mothers*, p. 36.
24. Thomas Denman, *An Introduction to the Practice of Midwifery*, 2nd ed. (London: J. Johnson, 1801), vol. 2, pp. 434–5.
25. Robert Gooch, *On Some of the Most Important Diseases Peculiar to Women; with Other Papers*, with a prefatory essay by Robert Ferguson, MD. (London: The New Sydenham Society, 1831), p. 54.
26. Thomas Denman included lengthy sections on mental disorders associated with childbearing in his treatises on midwifery published in 1801 and 1810; see Marland, *Dangerous Motherhood*, pp. 14–15, and for other descriptions of mental disturbance associated with childbirth prior to 1800, pp. 12–14.
27. George Man Burrows, *Commentaries*, pp. 363–4.
28. See Anne Digby, 'Women's Biological Straitjacket', in Susan Mendes and Jane Rendall (eds), *Sexuality and Subordination: Interdisciplinary Studies of Gender in the Nineteenth Century* (London and New York: Routledge, 1989), pp. 192–220, esp. p. 197.

29. William Smellie, *A Treatise on the Theory and Practice of Midwifery*, 3rd ed. (London: D. Wilson and T. Durham, 1756), vol. I, pp. 395–6.
30. See Thomas Dixon, *From Passions to Emotions: The Creation of a Secular Psychological Category* (Cambridge: Cambridge University Press, 2003) for the displacement of differentiated typologies, including passions, appetites, affections and sentiments, by an overarching category of emotions during the nineteenth century. See also Dixon's essay in this volume.
31. John Charles Bucknill and Daniel Hack Tuke, *A Manual of Psychological Medicine* (New York and London: Hafner, 1968 [1858]), pp. 235–6.
32. Reid, 'On the Causes, Symptoms, and Treatment', p. 141.
33. I. Loudon, 'Puerperal Insanity in the 19th Century', *Journal of the Royal Society of Medicine* (1988), 81, pp. 76–9, 76.
34. Bucknill and Tuke, *A Manual of Psychological Medicine*, pp. 235–9.
35. See Anne Digby, *Making a Medical Living: Doctors and Patients in the English Market for Medicine*, 1720–1911 (Cambridge: Cambridge University Press, 1994), chapter 9.
36. See e.g. Donnison, *Midwives and Medical Men*; Ornella Moscucci, *The Science of Woman: Gynaecology and Gender in England 1800–1929* (Cambridge: Cambridge University Press, 1990), chapter 2. The essays in Hilary Marland and Anne-Marie Rafferty (eds), *Midwives, Society and Childbirth: Debates and Controversies in the Modern Period* (London and New York: Routledge, 1997), seek to provide a more complex analysis of the male takeover of midwifery practice.
37. Fleetwood Churchill, *On the Diseases of Women; Including those of Pregnancy and Childbed*, 4th ed. (Dublin: Fannin and Co., 1857), p. 737.
38. Burrows, *Commentaries*, pp. 362–3.
39. For more details on treatment regimes, and the combination of moral and physical approaches, see Marland, *Dangerous Motherhood*, esp. pp. 44–8.
40. See ibid., chapter 2.
41. Dixon, *From Passions to Emotions*.
42. Edinburgh University Library: Lothian Health Board Archive, Royal Edinburgh Hospital, Case Book: LHB7/51/6, 1847–51, Jane Anderson or Rutherford, admitted 6 October 1848, p. 244.
43. Ibid., LHB7/51/15, 1862–65, Agnes Watson, admitted 1 May 1865, p. 704. See Marland, *Dangerous Motherhood*, chapter 4 for numerous examples of the florid behaviour of Edinburgh Asylum patients as reported in the casebooks.
44. James Reid, 'On the Causes, Symptoms and Treatment', p. 143.
45. Robert Gooch, *A Practical Compendium of Midwifery; Being the Course of Lectures on Midwifery, and on the Diseases of Women and Infants*, Delivered at St. Bartholomew's Hospital, by the Late Robert Gooch, MD. (London: Longman, Rees, Orme, Brown, and Greene, 1831), p. 294.
46. Robert Boyd, 'Observations on Puerperal Insanity', *Journal of Mental Science* (1870), 16, pp. 153–65, 156. See Marland, *Dangerous Motherhood*, pp. 160–5 for views on the role of husbands in provoking puerperal insanity.
47. Gooch, *A Practical Compendium of Midwifery*, p. 290.
48. Gooch, *On Some of the Most Important Diseases Peculiar to Women*, p. 63.
49. John Burns, *The Principles of Midwifery*, 3rd ed. (London: Longman, Hurst, Rees, Orme, and Brown, 1814), pp. 434–5.
50. Marshall Hall, *Commentaries on Some of the More Important of the Diseases of Females* (London: Longman, Rees, Orme, Brown and Green, 1827), pp. 251–2.

51. W. E. Image, 'Case of Melancholia Puerperalis Attonia', *London Medical Gazette* (1845), 1, p. 281.
52. James Y. Simpson, *Clinical Lectures on the Diseases of Women* (Edinburgh: Adam and Charles Black, 1872), pp. 566–7.
53. J. T. Conquest, *Outlines of Midwifery, Developing Its Principles and Practice;* Intended as a Text Book for Students and a Book of Reference for Junior Practitioners (London: John Anderson, 1820), p. 229.
54. Gooch, *On Some of the Most Important Diseases Peculiar to Women,* p. 63.
55. Ibid., p. 77; Gooch, *A Practical Compendium of Midwifery,* p. 290.
56. Warwick County Record Office: Warwick County Lunatic Asylum, CR 2379/3, Case Book, 2 April 1864–13 May 1867, Harriet Ashmore, admitted 26 May 1864 (1864).
57. Ibid., Emma Walker, admitted 8 November 1864 (1259).
58. John Batty Tuke, 'On the Statistics of Puerperal Insanity as Observed in the Royal Edinburgh Asylum, Morningside', *Edinburgh Medical Journal* (1864–65), 10, pp. 1013–28, 1015.
59. William C. Ellis, *A Treatise on the Nature, Symptoms, Causes, and Treatment of Insanity, with Practical Observations on Lunatic Asylums, and a Description of the Pauper Lunatic Asylum for the County of Middlesex, at Hanwell* (London: Samuel Holdsworth, 1838), p. 241.
60. See Marland, *Dangerous Motherhood,* chapter 3 for the relationship between puerperal insanity and the bourgeois home.
61. Barbara H. Rosenwein, 'Worrying about Emotions in History', Review Essay, *American Historical Review* (2002), 107, pp. 821–45.
62. Ibid., p. 842.
63. Ibid., p. 827. See Norbert Elias, *The Civilizing Process: The History of Manners and State Formation and Civilization,* trans. Edmund Jephcott, 2 vols (Oxford: Blackwell, 1978, 1982).
64. Erving Goffman, *Asylums: Essays on the Social Situation of Mental Patients and other Inmates* (New York: Doubleday Anchor, 1961); Michel Foucault, *Madness and Civilization: A History of Insanity in the Age of Reason* (London: Tavistock, 1967).
65. See chapter 4 of Marland, *Dangerous Motherhood* for the experiences of patients in the Royal Edinburgh Asylum in the mid-nineteenth century.
66. John Conolly, *An Inquiry Concerning the Indications of Insanity, with Suggestions for the Better Protection and Care of the Insane* (London: John Taylor, 1830) p. 428.
67. John Conolly, 'Description and Treatment of Puerperal Insanity', Lecture XIII: 'Clinical Lectures on the Principle Forms of Insanity, Delivered in the Middlesex Lunatic-Asylum at Hanwell', *Lancet* (28 March 1846), I, pp. 349–54, 351.
68. For more on Gooch and selected case notes, see Marland, *Dangerous Motherhood,* esp. pp. 29–31, 71–2, 74–7.
69. Gooch, *On Some of the Most Important Diseases Peculiar to Women,* pp. 56–7.
70. Arthur H. M'Clintock and Samuel L. Hardy, *Practical Observations on Midwifery, and the Diseases Incident to the Puerperal State* (Dublin: Hodges and Smith, 1848), p. 67.
71. R. U. West, 'Fatal and Other Cases of Puerperal Mania', *Association Medical Journal* (11 Aug. 1854), 2, pp. 716–18, quote on p. 718.

72. Ibid.
73. Ibid.
74. Gooch, *A Practical Compendium of Midwifery*, p. 292.
75. John Gregory, *Lectures on the Duties and Qualifications of a Physician, Revised and Corrected by James Gregory,* MD (Edinburgh and London: William Creech, T. Cadell and W. Davies, 1805), p. 17. Cited in Dixon's essay in this volume.
76. M'Clintock and Hardy, *Practical Observations on Midwifery*, p. 68.
77. William Tyler Smith, 'Puerperal Mania', Lecture XXXIX: 'Lectures on the Theory and Practice of Obstetrics', *Lancet* (18 October 1856), II, pp. 423–5, 424. See also Dixon's essay in this volume on the role of the physician in managing the passions.
78. Bucknill and Tuke, *A Manual of Psychological Medicine*, p. 88. See also Laurence J. Ray, 'Models of Madness in Victorian Asylum Practice, *Archives of European Sociology* (1981), 22, pp. 229–64, for the stress model of early nineteenth-century alienists.
79. Robert Gooch, *Observations on Puerperal Insanity* (London, 1820) (extracted from sixth volume of *Medical Transactions*, Royal College of Physicians, read at the College, 16 December 1819), pp. 4–6.
80. The question of how this changed towards the end of the century is addressed in Marland, *Dangerous Motherhood*, Chapter 7.

4

'Cold Calculation in the Faces of Horrors?' Pity, Compassion and the Making of Humanitarian Protocols.[1]

Bertrand Taithe

The histories of compassion and humanitarianism go together hand in hand. Humanitarian agencies since the First World War have called in a very public manner on international compassion. This chapter will argue that the Red Cross, since its intellectual foundation, has also called on a language of emotions, in this particular instance compassion, in order to obtain the financial and political resources required by an international organization. Similarly, the multitudes of agencies dealing with disasters today compete, with imagery and texts, to obtain a response, materialized as a gift. Their association with compassion has tended to reify the concept of compassion as a given, casting their practices as a constant devoid of history. Yet the history of compassion as an emotion, which became socialized through institutions, is not without its complexities and challenges. This chapter intends to explore these complexities by considering in turn the ethical dilemma that compassion presents, and the historical context of the making of compassionate protocols.

Pity and compassion have a discrete existence in the nosology of emotions. To the ancients, Konstan argues, pity was essentially an emotion produced by a 'cognitive assessment' overriding the initial repulsion or desire to avoid the undeserved misfortune of others.[2] The theme of pity, later Christianized and modified as compassion or sympathy (feeling *about* another person), became a central theme in eighteenth-century philosophy.[3] Sympathy and pity (including the paradoxical self-pity) were central to any discussion of morality; their absence could signal a pathological condition. Yet an excess of either could also be pathological.[4] Investigation of the emotions since Charles Darwin's *Expression of the Emotions in Man and Animals* (1872) has subsequently influenced the development of modern psychology.[5] One could attempt to pursue Wierzbicka and Konstan's admirable studies of the comparative semantic and cultural meanings of pity and compassion from an individual psychological or philosophical point of view, but this chapter is devoted to a different set of questions.[6] It is primarily concerned with the manner in which the cognitive assessment of an individual's emotion could become that of a group, and how compassion and humanitarian feeling could become 'organized' and structured in the modern era under the label of 'humanitarianism', particularly as applied to medical practice.[7]

In some ways this chapter deals with the waxing and waning of linguistic constructions of emotions and their embodiment in social formations. Undoubtedly 'pity' is now perceived as condescending, while the 'pitiful' (*pitoyable* in French) is generally an object of dismay and contempt. Similarly, general discussions of pity, compassion and charity have been framed in radical political terms. Feminist scholars and disabled rights militants and scholars have described pity as the expression of 'colonizing appropriation' of the plight of others or even as the inherently conservative and un-reasoned reinforcement of the social status quo. In this sense, pity-like charity has been tainted as a patronizing emotion, hence the ironic comments at the expense of recent evocations of the 'compassionate conservatism' of the Republican Party in the USA.[8] Yet, as this chapter demonstrates, the term 'humanitarian/-ism' has followed a reverse journey; from being a derisory term, this is now a contested but largely revered mode of feeling about the world. Exploring a moment during the mid-nineteenth century, it will show how compassion and pity were re-cast according to a different set of social obligations, and when these emotions, framed by a humanitarian narrative, became subject to the compassionate protocols of new organizations such as the Red Cross. The main focus here is the Francophone world, from which most of the sources used here originate, but the argument applies elsewhere in Western Europe. Indeed, a key feature of humanitarianism is that it claims its international dimensions as a form of universalism. Though this is only a fraction of the debate on compassion, it is one that has gained a real centrality in our culture.

The golden age of humanitarian compassion was initially conceived as the late Enlightenment. In 1989 Thomas Laqueur suggested a number of central features specific to eighteenth-century humanitarian textual sensibilities.[9] The central focus of his argument was how narrative fostered the arousal of compassion and the transformation of compassion into a moral imperative. Compassion arose, according to Laqueur, from the myriad detailed narratives of the body in pain, which he then characterized as 'the humanitarian narrative'.[10] According to Laqueur, empathy alone (feeling *with* another person) explained how these narratives could provoke a compassionate response in the reader. This empathy literally enabled the appropriation of the suffering of others. The varied sources Laqueur used included early novels, accounts of autopsy and clinical writings as well as reformist discourses. In all these texts, linked by their common structural narrative devices, he assumed the positive nature of compassion and empathy. Stressing the need to contextualize over a period of one hundred years, Laqueur added in a caveat that by the - mid-nineteenth century humanitarianism had become a term of contempt for the sentimental attachment to distant sufferings at the expense of closer ones. Beyond Mary Poovey, who picked up the narratives of reform in a broad Foucaultian framework, few scholars have revisited the history of humanitarianism beyond the confines of the anti-slavery struggle begun in the late eighteenth century.[11] The actual impact of the humanitarian rhetoric of sentimentality in relation to the hard facts of the declining slave economy has led to a debate which measured, quite rightly, the distance between formulaic compassion and policy-making.[12] Contemporaries reading about slavery and yet helping themselves to 'blood-stained cane sugar' arguably betrayed the same inconsequential approach that characterizes many modern responses to humanitarian messages and the general inconvenience of adopting an ethically consistent lifestyle.[13]

Revisiting the humanitarian debate along the linguistic turn tropes set up in the 1980s, Karen Haltunen has explored the dark side of empathy and compassion, and evoked directly the pornography of pain in these texts.[14] The Latitudinarian divines pushed to extremes the moderate and enlightened forms of empathy evoked by Laqueur in their literature of excess. The gross representation of pain did not necessarily connect the body of the other to one's own in an empathetic fashion, but instead derived some of its power of attraction from the desire of objectification. Thus the dark eroticism of Sadeian sexuality was also a product of the Enlightenment. Coming from a radically different perspective, French ethicists and clinical specialists of pain have come to similar conclusions, calling in the French philosopher Philippe Sollers to denounce the ambiguities of enlightenment empathy with sufferings.[15]

This chapter argues that the history of modern humanitarian compassion is neither linear nor a civilizing process. Its nineteenth-century journey is one of transformation into organized and codified forms, and also contests, because of its new organization and because of deeper questionings of the nature of compassion, and of compassion's relationship with other figurative modes of expression of interest in the other: love, concern, care. Many of these have pre-Enlightenment forms or have been taken over by responses to the Enlightenment. This chapter considers in turn the philosophical re-casting of pity and compassion in the nineteenth century. At its heart it considers how Henry Dunant's exemplary *A Memory of Solferino* set to develop the cognitive processes of an emotional narrative, building on religious tropes of sacrifice, into compassionate protocols of organized humanitarianism.[16]

Empathy and humanity

The key process whereby a witness can transcribe his or her feelings into a cogent narrative, which would then provoke, through the power of words and images, an empathetic response in readers, presents all the central problems of literary and historical analyses of texts. The empathy thus described is one of the key conceits of bourgeois narratives, one of the 'functions of texts', which most literary critics from Walter Benjamin to Marthe Robert or Roland Barthes have denounced as a convenient fiction and an interpretative straitjacket. Texts are not read the way they were meant to be read; the empathy they may generate can also misfire. When Sade subverted the genre of conventional gothic novels to create a manic enumeration of vices and sufferings, he clearly reflected on the arousal that beautiful descriptions of pain might provoke among his readers. Reversing the propositions of more conventional licentiousness by pornographers like Retif de la Bretonne, who emphasized lust and pleasure, Sade's emphasis was on descriptions of pain as pleasure or even on the spectacle of the death or subjection of the other as orgasmic.[17] The description of pain might be arousing but not necessarily of compassion. It can also be ineffectual and lead to saturation and tedium. Suffering fatigue is indeed not only a recent concern of humanitarian agencies, but is intrinsic in all narratives of pain. The French Revolution itself had demonstrated the limits of empathy. Retif de la Bretonne, walking the streets of Paris during the Terror in his *nuits révolutionnaires,* mused on the fragility of that process. Corpses and tragic stories abounded in 1792–1793, to the point of arousing nothing but occasional fear and indifference. Yet one could still feel some sympathy for the fate of the king and the royal family; to the dead an absence of feelings, to the condemned an emotion, real yet circumscribed by the political certitude of their guilt.[18]

Constantly evoking the language of emotions and sensibility, Retif wondered on compassion in moral and historical terms:

> To relieve myself [from emotions] I was drifting in the following centuries. I saw the men of 1992 reading our history. I sought to hear them and hear them I did. The severity of their judgement scared me. It seemed to me that some were reproaching our lack of humanity while the extremes, such as they are today, alone approved us.[19]

Nineteenth-century historians from Lamartine to Michelet later attributed to this emotional response the decline of republican ideals and the monarchical restorations of 1814 and 1815.[20] The French Revolution challenged humanitarian modes of narration by the lavishness of its violence and the public enjoyment of suffering from the September massacres to the regular executions. Arguably the Enlightenment already contained the seeds of this challenge to any consensus on the human condition and its philosophical status. Carlo Ginzburg has paid particular attention to Diderot's *Letter on the Blind* (1749), which posited the solipsistic indifference of the disabled unable to distinguish between 'a man urinating and a man bleeding to death'.[21] In Ginzburg's view, this text illustrated the dilution of unmediated compassion by distance (unable to see, the blind man would be as remote as others are to the sufferings of the distant and unknown men). Another reading informed by disability studies would re-cast Diderot's perspective somewhat. While the metaphor of geographical distance remains, and will be discussed further below, it also affects directly the objects of compassion nearby. In Diderot if a blind person is unable to reciprocate compassion as a result of the disability damaging his or her cognitive appreciation of others, what sort of unilateral compassion should apply to him or her? What humanity is there to share with any solipsistic being? This philosophical distance is not merely spatial. The narration of disablement and suffering established another distance and an irreducible form of alterity, which in turn diminished the more abstract common bind of humanity at the heart of compassion.[22] This reflection on compassion could of course be contrasted with Adam Smith who distinguished sympathy (a universal feeling) with compassion, the 'fellow feeling' towards the sorrow of others.[23]

From Schopenhauer's meditation on the power of compassion in 1818 to Nietzsche, nineteenth-century philosophers have continually ruminated on this conundrum to the point of casting a doubt on the ability, or necessity, of empathy. Kant famously regarded pity as potentially pathological and as the source of increased suffering. Nietzsche furthermore argued that the danger to the 'pitier' extended to that of becoming a victim of their object of pity. In this sense empathy enabled the sufferers eliciting pity to exercise their own power to hurt the 'pitier'.[24] The power relation was of course reversible in a less counter-intuitive manner for Nietzsche since 'pitiers' could seek objects of pity either to aggrandize themselves or to enslave others in binds of exclusive dependency.[25] In contrast to this bleak power analysis of relationships based on pity, Schopenhauer posited compassion as the origin of all virtues in a bleak world in which only self-observation could lead to an understanding of nature.[26] His early-nineteenth-century pessimism also redefined pain as the positive expression of life while pleasure became merely its absence.[27] In spite of his insistence on compassion, a solipsism of being remained as the inability to fully engage with the other, without shaping the other as another self or a mirror of opposites.

While very schematic, this philosophical framework enables us to comprehend the nineteenth-century political polarities of pity, compassion and humanitarian feelings.

If the above discussion of empathy seems a little abstract, there is ample literary evidence to show that the objects of pity and compassion elicited complex emotional responses. Many nineteenth-century writers who did not fall for a melodramatic mode of representation of illness or disability noted the growing indifference of people nearest to a patient suffering from long-term ailments. In this case the repetitious nature of the patient's complaint usually explained the indifference of their kin. Compassion or empathy in this process did not come naturally. Even if Harriet Martineau argued in *Life in the Sick-Room* (1844) that consoling came naturally to the simple-minded people, their simple-mindedness would in itself be an obstacle to empathy in situations they could not envision.[28] As Martineau pointed out again, the common perception of pain in others is that they (the sufferers) can become inured to it.[29] Then pain becomes the essence of their being, and, as a result, it is just another form of the human condition and one that does not necessarily stimulate compassion. With this complexity in mind, humanitarian narratives can never by themselves arouse compassion and texts do not have the power to elicit emotions, any more than emotions are historical agents in themselves. To most commentators, meliorists as well as cynics, the idea that humanitarian texts could produce change or good by themselves seemed debatable or even vacuous.

On one extreme of the social commentaries in the 1830s, the descriptions of *humanitarianism* as excessive sensibility, in French *sensiblerie* (tearful sensitivity), can be found in Balzac who clearly identified the limitations of humanitarianism. Balzac coined the term in 1837 in a novel, *Les Employés,* set in 1824 in which he attacked humanitarianism as a high level of abstraction devoid of real intensity and individual compassion through his description of a republican character:

> His heart was full of this stupid collective love that we have to name *humanitarianism*, elder son of defunct philanthropy, and which bears the same relation to divine Catholic charity than a system to Art, than Reasoning substituted to a masterpiece [*Oeuvre*].[30]

Balzac, who indisputably shaped the French novel of the first half of the nineteenth century, focussed his analysis of compassion on the moral debates arising from the Enlightenment, and in particular from the themes arising in Diderot and Rousseau, and on the rise of ruthless materialism.[31] One of the key literary moments of any discussion of the limitations of compassion is in the famous metaphor of the death of the Chinese mandarin.[32] In *Le Père Goriot,* set in 1819, Vautrin, the incarnation of amoral ambition, presents to the hero the opportunity to enrich himself by wishing the death of an unknown man.[33] This precise metaphor was loosely inspired by Diderot (although Balzac misattributed it to Rousseau), to illustrate the universality of morality: in eighteenth-century moral philosophy distance should not abolish moral duty. Compassion should not be diluted in space and the old Chinese mandarin might live for the sake of humanity. In Balzac the hero's friend casually declares 'I am at my thirty third mandarin'. Later on the hero attempts, unsuccessfully, to prevent the unknown man's death for which he had fleetingly wished. The unknown man nevertheless dies in a set-up duel. Balzac's metaphor is clear: the Chinese mandarin has to die for modern wealth to be generated and nothing can be done to save him. The desires to gain material wealth matter more to the modern man than any abstract sympathy for unknown men. Vautrin, knowingly reversing every trope of eighteenth-century humanitarianism, then intends to retire with his share of the proceedings on a slave plantation ('I want two hundred Negroes in order to satisfy my taste for a patriarchal lifestyle') in a southern state of the U.S.A.[34]

In Balzac's work, Vautrin recurs often as a prototype of modern man and the archetype of the age. He stands opposed to any form of weakness such as the romantic feminine characters or to the elderly character of Père Goriot. Modelled on Eugène Vidocq, the repented criminal in charge of the police *agents provocateurs* from 1812,[35] he demonstrates how blurred the moral boundaries are in the age of materialism. He incarnates strength and resolve in a luciferian manner but his character stands as the critique of vacuous romantic empathy or abstract humanitarianism. This critical perspective on abstract humanitarianism endured throughout the nineteenth century, and until the First World War other critiques and cynical commentators denounced the vacuity of humanitarianism in equally scathing terms: 'humanitarianism is a special form of *sensiblerie* reserved for educated people, based on philosophical, social and philosophical abstractions from which it extracts the quintessential [forms]'.[36]

Mediating sufferings: Henry Dunant and humanitarian narratives

While a Balzacian sensibility, materialistic perspectives, and post-revolutionary doubts with systematic abstractions endured, the second half of the nineteenth century nevertheless witnessed new paradigms of compassion in the shape of new organizations devoted to the cognitive assessment of the suffering of others and new narrative forms aiming at mediating compassion and love. In this sense, the key texts of the nineteenth century are Florence Nightingale's letters from Crimea and Henry Dunant's account of the battle of Solferino.[37] The latter will be the particular subject explored in this section.

Henry Dunant, a Swiss businessman whose insolvent Algerian venture required further land concessions from the French government, set on finding Napoleon III on the battlefront of Italy.[38]As pointed out in the hagiography of Dunant, he was inspired by fervent Christian idealism even though his original motives were primarily economic.[39] By accident, Dunant found himself near the battlefield of Solferino where he stayed a few days. The experience led him to write *A Memory of Solferino*, which he originally printed at his own expense. This he sent to influential figures throughout Europe, including Gustave Moynier, president of the Society for Public Utility in Geneva and later the real organizer of the Red Cross movement. The book had a long career and numerous reprints throughout the nineteenth century, but its real impact was to be measured at the legislative level since the conference of 1863, and the Geneva Convention of 1864, based on this book, can legitimately be described as the cornerstone of modern internal humanitarian law.[40] Ultimately, at the end of a life of subsequent bankruptcy and wilful obscurity, the book obtained the first Nobel Prize for Peace in 1901 for its author.

From the perspective of this exploration of the history of compassion and emotions, the text of Henry Dunant stands out as an exemplary engagement with the difficulties of transcending indifference, and renewing the structure and tropes of the humanitarian narrative. Compared with the texts produced by slavery abolitionists, for instance, Dunant had to break away from the individualism of compassion and its single purpose (the abolition of slavery in one place, at one time), to redefine the parameters of a new universalism embarking on an open-ended quest. War will not end, and cannot be abolished, from Dunant's perspective. Dunant's 1862 account of the battle of Solferino instead highlighted what compassion might achieve, were sentiments and vague feelings of revulsion converted into positive action. Yet Dunant did not spare his readers and his narrative uses some essential tropes of narratives of details. He described in details the saws grinding into the bones of the amputees; he transcribed their cries in pathetic melodramatic tones. He painted a vivid picture of an infernal ambulance where one casualty was spilling his brain on the floor while others, looking aghast, disappeared under a swarm of flies. This relatively short book is replete with a mixture of conventional accounts of the attack on the village of Solferino by the Franco-Sardinian forces, and the minutia of individual pains and sufferings. Anonymous soldiers and named officers, following war narrative conventions, receive equal attention. In pure narrative terms, Dunant is self-conscious of his transgressive attention to shocking effects, but he also offers a way out by emphasizing action over emotion:

> [T]he moral sense of human life; the humane desire to lighten a little the torments of all these poor miserable wretches, or restore their shattered courage; the furious and relentless activity which a man summons up at such moments: all these combine to create a kind of energy which gives a positive craving to relieve as many as one can. There is no more grieving at the multiple scenes of the fearful and solemn tragedy. There is indifference as one passes even before the more frightfully disfigured corpses. There is something akin to cold calculation in the faces of horrors yet more ghastly than those here described and which the pen absolutely declines to set down.[41]

This calculation is here a *positive* engagement and response detached from the devastating emotionality of excessive sensitivity.[42] Revealingly, the descriptive texts that originated in the Red Cross movement in the ten years following this seminal book concentrated their energy on the calculated accounts of suffering rather than on emotive descriptive material, leaving this specific transitional role to the founding narrative of *A Memory of Solferino*.

This central use of *A Memory of Solferino* was not merely to mediate compassion from the eyewitness to the reader but to ensure that this mediation was taken further away from the realm of personal emotions, potentially damaging to all concerned, into the domain of stimulus-social response. Between Dunant in 1862 and now, the humanitarian narrative developed further by confronting modern philosophical and medical challenges questioning the meanings of compassion and pain, and religious approaches to compassion. Structurally, its main innovation was to escape from the pure literary field to become a narrative form leading to the objectification of suffering and to the 'cold calculation' of effective compassionate action. In narrative terms, another essential shift has been from a monocausal campaign, the abolition of slavery which would end when its ambition was fulfilled, to an essentially open-ended narrative which is both extremely abstract and yet absolutely practical: humanitarian interventionism. The abolition of pain is not foreseeable, but the time- and space-limited intervention intending to limit pain, in one place, for a period of time, is. In Dunant the action proposed, a new society devoted to alleviating but not suppressing the horrors of war, was an alternative to the narratives of individual fates that defy description. It was a revised form of compassion, and indeed one that would take the shape of an organization, with a structure, an ethos and a legitimacy backed by international law.[43] Dunant thus added in a footnote: 'if these pages could bring up the question (or lead to it being developed and its urgency realized) of the help to be given to the wounded soldiers in wartime'.[44] In this sense, Dunant's narrative of 1862 redefined the nature and purpose of humanitarian narratives and the meanings of compassion in an age of international exchanges and of the globalization of compassion. He did so not only by establishing the modalities of the application of institutional compassion, but also by developing the parallels between the humanitarian narrative and the open-ended story of modernity.[45]

In this attempt to mediate emotional responses and channel them into a discourse of action, Dunant was not alone. A contemporary of Florence Nightingale, Dr Chenu, had long developed an analysis of the causes of mortality by neglect and had developed a discursive strategy based on the calculation of wasted human lives.[46] By contrast with Dunant, Chenu used the rhetoric of numbers, the language of demographers and statisticians, dominant among policy makers in France.[47] His approach had been noted and well reviewed in most influential papers, but ultimately failed to raise wider support – unlike Dunant whose prose appealed equally to policy makers and the public.[48] Military sources criticized this neutral language and particularly responded negatively to its dispassionate tone – the lack of emotion and the calculating features of his rhetoric.[49] What made Dunant successful in his reinvention of the humanitarian narrative was his ability to mix pathos and heroics with calls to organizations and anecdotes, and to mingle the imagery of glory with the full range of emotional responses to suffering developed in contemporary religious discourses on compassion, and, in France, particularly in the Catholic Church.

Compassion, pain and religion

The Catholic Church, and in general religiosity embracing mysticism, presented a key alternative mode of humanitarian narration that owed little to the Enlightenment except the thorough rejection of its objectives and assumptions. It offered an alternative narrative of compassion to the philosophical discussion mapped, however rapidly, in the first section. The use of shocking accounts in new humanitarian narratives derived from this perception that compassion is indeed a most obscure reflex in the human psyche, and that the rational and sentimental philosophical accounts of the eighteenth century inadequately explored these recesses. Turning to alternative accounts was logical for disenchanted authors such as Joris-Karl Huysmans whose *Against Nature* signalled a fin-de-siècle anxiety with solipsism and narrated the fate of a man unable to engage emotionally with others while facing his own mental and physical sufferings.[50] Joris-Karl Huysmans then turned away from the naturalism of his young writing career to the fold of the Catholic Church and became a devout Christian, leaning towards the most explicitly anti-modern aspects of Catholic theology. Within the Catholic ethos the status of victim plays a central role in narratives of redemption. The adoration of the crucified and martyrs was at the heart of the theological counter-Enlightenment crusade of the mid-nineteenth century. The famous pastoral of love involved the proliferation of vivid images of suffering.[51] In France the *Saint-Sulpice* Paris seminary, which produced eminent theologians such as Renan or Lavigerie, gave its name to a genre of realistic sculptures and paintings, dripping with the blood of the saints, which were mass-produced and distributed throughout the land.

For the martyrs, even in the midst of their agony, ecstasy was often portrayed vividly as their reward. Richard Burton has recently made the point that this theology has had a much deeper impact than is often assumed.[52] Similarly Christopher Forth and Ivan Strenski have highlighted the pervasiveness of cults of sacrifice that went as far as the active pursuit of martyrdom.[53] Among nuns recruited in times of emergency to serve as nurses, the acceptation of pain and disease in that spirit went as far as the rejection of vaccination in the great small pox epidemic of 1870.[54] The enviable status of the victim, often female, reversed much of the logic of the Enlightenment humanitarian narrative. It proceeds notably by elevating the abjection of pain or suffering to an offering to God. From Catherine of Sienna's writing, which described vividly a desire to absorb physically and mentally the sufferings of the sick, to the Theresa of Lisieux at the end of the nineteenth century who elevated her agony to an ecstatic progress towards the divine, we have numerous instances of accounts relishing the sufferings of others and self. The Counter-Reformation figure of Catherine of Sienna was notoriously at the heart of much attention from the Protestant feminist campaigner Josephine Butler to the atheistic psychiatrist Charcot; her sensibility resonated in a very striking way with both the humanitarian enthusiasm of Butler and the psychological investigations of Charcot.[55] In Lourdes, where the faithful and desperate cases congregated in the hope of miracle, Catholic enthusiasts went as far as drinking water contaminated by pus and blood in communion with the expiatory pains of the sick.[56] Indeed this emphasis on the beauty and troubled desirability of abjection arose simultaneously in the new humanitarian campaigns of the late nineteenth century when Mgr Lavigerie, Archbishop of Algiers, sought funds for the malnourished orphans of the great Algerian famine by stressing their readiness to feed themselves with excrement and corpses.[57] Similar tropes of abjection and exploitation, whereby the object of charity was simultaneously repulsive and corrupted and a helpless victim, were regularly deployed in Western Europe for the converted sinner or the fallen.[58]

Most recently the theme of 'abjection' in Queer studies has become an avenue to explore the boundaries of self-identity confronted by desire and love. The theme can be found in the 1930's works of the Catholic writer Jouhandeau or the 1950's writings of Jean Genet, but since Kristeva's work, it has been used increasingly to describe the more unspoken nexus of gay sexual identity and the self-destructive fascinations of love. The abject is thus intrinsically what is within and without, what constitutes and threatens self-perception.[59] If the body abject is not only the object of pity but also of desire in its debased form, the meanings of compassion are altered and compassion can indeed become a quest for troubling objects of desire. The idea that the humanitarian narrative might be seeking to describe pain in others, not as a terrifying anomaly, but as an object of emulation or even desire, may seem unsettling. Yet it helps understand the often obsessive 'will to know' that characterizes humanitarian narratives. The will to uncover pain in its many shapes to bring it home is not *only* the result of a 'pure' moral imperative, but is essentially a development of the cognitive processes of emotion, a form of dialectic empathy which did not necessarily lead to a desire to abolish the causes of abjection. Crucial to this approach, which transcends meek understandings of empathy, is the ontological status of pain and suffering in medical and humanitarian accounts.

To move from the concept of sacrifice and martyrdom to that of pain is not necessarily as straightforward as one might think. The ecstasy of extreme pain in sacrifice and martyrdom sprung from the reward sought in heaven and the very literal imitation of Christ. Yet from the mid-nineteenth century onwards, medical reflection on the nature of pain changed from a purely symptomatic reading to a neurological one. Following from the work of Roselyne Rey,[60] it is clear that the real epistemological break took place at the turn of the twentieth century. From this perspective the concept of psychic suffering, partially rooted in psycho-somatic interpretation of mental illnesses, paved the way for a mutual deciphering of suffering which could leave its imprint on both the sufferer and the witness in the form of physical and psychological trauma.[61] Even in the early premises of psychological analyses of the consequences of attending the grotesquely ill, eyewitnesses often commented on the dangers inherent in facing extremes of pain in others. Henry Dunant, surrounded by willing tourists, journalists, priests and officers in his impossible rescue work in Castiglione Della Pieve, near Solferino, stated the two dangers of breakdown or feverish exaltation:

> [B]efore long one of the latter [officers] found that the scene made him ill, and our other volunteers withdrew one by one, for they could no longer bear to look upon suffering which they could do so little to relieve... [for another volunteer] it became necessary, for his own sake, to restrain his own zeal, and we had also to calm the sympathetic excitement of a Belgian, which had become such that we feared he might have an access of high fever.[62]

Here the pain could still appear by osmosis in the eyewitness, unhinging him to the point of breakdown. The processes of empathy, emotion and compassion were dangerous things indeed when they could so much affect the observer that they might blur the boundaries between the sufferings of the observed and the observer.

Compassion and painkillers could nevertheless go together, and adequately attenuate the more violent scenes of ecstatic agony, without necessarily removing their aesthetic appeal. Of course anaesthetics existed in the days of Dunant but it is characteristic that none is used in his narratives and that he preferred descriptions of an 'ecstasy of pain'.[63] The trance-like attitudes of ecstasy were communicative in religious settings but this level of communion was not to be allowed among humanitarian workers. The modern science of pain would increasingly enable procedural boundaries to prevent the disabling effects of excessive empathy. In the heroic days of medical experimentation on the origin of pain, the French colonial doctor Jules Regnaut, who operated on his own hernia in 1912 under local anaesthetic, thus claimed to have played on his own nerves 'as on a mandolin', identifying the exact point were anaesthetics failed to operate before furthering his work on the uses of pain (including torture) and its cultural variability among non-Western societies.[64] The Catholic Church, in the years following the First World War, sought to engage in a similar dialogue with medical understanding of pain. In France, at least, the Church was deeply involved, through its religious orders of nun-nurses, in the therapeutic treatments of pain and its teaching, together with analgesics, was delivered in most hospitals to the terminally ill. In this sense the religious imagery and the spectacle of medical distance were a diptych offered to the ill. To the terminally ill, religion offered the recourse of a compassionate hearing, a sacralization of the fatal sufferings or even the hope of a miracle.

In matters of dogma, from 1883 the church had engaged medical expertise in the analysis of miracles and saintly events, thus modernizing its multi-secular hermeneutic of suspicion towards unnatural events.[65] Beyond the miracles of Lourdes, mystics exhibiting stigmata, who pushed the imitation of Christ to the point of presenting his crucifixion wounds to the hands, side and head, presented a particular problem unique to Western Christianity.[66] In the journal of *Études Carmélitaines* of 1936, doctors and philosophers were invited to discuss critically of the nature of pain and how it should be considered as well as give their, usually sceptical, verdict on recently observed stigmata. On pain the religious orthodoxy held fast and promoted its value and ontological status:

> [I]if pain is taken from the side of life, which is enduring, and not from the side of the illness [also evil: le mal] which is undermining life, suffering will appear more like life in pain and I wonder what could be more positive on earth than that. Life becomes more intense where it is wounded and hindered, than where, being unchallenged, nothing provokes it to shout its existence. A superior living being will be able to experience more nuances and a richer range of suffering.[67]

The development of the humanitarian narrative cannot be separated from its religious sources and the emotional response to pain. The extremes, where medical treatments could only alleviate temporarily, were precisely the places where religion and medicine could interact best. If the meanings of compassion raise some important questions on the implied psychological mechanics of an appropriation of the pain of others and of its purpose, the meanings of pain imply a transcendent approach to the other. In this sense the humanitarian narrative is cast as a reflection on the universalism of life, and the human condition, glorifying suffering as much as deploring it.

The victims' innocence is thus directly linked to their suffering, yet the relief they might obtain as a result of humanitarian action will have to proceed through an organizational filter, a dispassionate and cold calculation which alone can be efficient enough to be *real* compassion and which will emulate the 'double-edged pity' of the surgeon since antiquity.[68] Beyond the aesthetics of suffering mentioned above, another essential difference from the eighteenth-century humanitarian narrative described by Laqueur is in the conversion from a 'passion' into a protocol legitimated by its efficiency.

Conclusion

There is no closure in the humanitarian narrative, and the modern war Dunant described is still raging. The wound is always running and open. The atrocities of yesterday or yesteryear only prefigure those, inevitably greater, of tomorrow. While the humanitarian narrative remains a narrative of detail (this child needs your help now) it is also a collective one (this society needs help from this one) and a cultural one (this civilization is failing to emulate ours).[69] If on the one hand the humanitarian narrative tells a fundamental and palpable truth about the universality of human sufferings (with some cultural relativism between the West and the East or the North and South), on the other it also lies fundamentally about how the ensuing humanitarian action might operate.[70] Many humanitarian narratives based on eyewitness accounts were thus set in the immediacy of a crisis or the endemic physiological misery of a group. They testified of the urgency of the 'now': this child, that woman, are dying, and will be dead if you, the reader, do not do something about it; whatever is done by the emotional reader will not affect this woman, this child, but another, perhaps elsewhere or even in radically different circumstances. The narratives of action lead nowhere in particular. The inevitability of these bleak assessments ultimately mean that unlike the eighteenth-century novel or other bourgeois narrative forms, the humanitarian narrative is written in the epic mode, that of quests and inconclusive battles, precisely like Solferino. The modern humanitarian narrative is one that is devoted to the transition between problematic individual compassion, a singular cognitive process, and a more systematic, at least symbolically effective (displacing mountains of food, equipment, machines and images), socialized extrapolation, which alone could carry through the collective delegation of compassion. Again, Dunant was central to this shift:

> Why have I lingered with seeming complacency over lamentable pictures, tracing their details with what may appear desperate fidelity? It is a natural question. Perhaps I might answer with another: would it not be possible in time of peace and quiet to form relief societies for the purpose of having care given to the wounded in wartime by zealous, devoted and thoroughly qualified volunteers?[71]

There is in this final extract a fascinating transition from a reader to 'societies'. How expertise might aggregate and form such organizations is left unexplained and it is not the purpose of this chapter to show how humanitarian societies develop. Nor is it explained how pain-relief might dramatically modify the agony of the soldiers Dunant had collected. What his emotional response, his compassion, entailed on the day was primarily pragmatic: to bring some water, to move slightly the dying, to segregate the living from the dead. From a historical point of view his acts have often remained as the singular founding event of humanitarianism, yet it was his emotion and the literary forms that it took that enabled new laws, international conventions and organizations. Ironically compassion had to be channelled and turned into something '*akin* to cold calculation' to become real. *Akin* is here crucially signalling how emotion hid itself or was contained. The action of the Red Cross represented real compassion rather than an ephemeral literary construct. In this sense the historian is looking at the self-denied role of emotions, at the imprint in the sand.

There was something futile in Dunant's compassion. He admitted elsewhere that most wounded soldiers were lost and that he simply humanized, but also prolonged, their agony and suffering. The humanitarian protocol obeyed higher principles but mostly existed for its own sake. This open-ended narrative invited readers to become the spectators of the next Solferino, to join the tourists and journalists who surrounded Dunant, for compassionate protocols are not without spectators.

Notes

1. The original version of this paper was presented at the Anglo–Dutch–German workshop in the history of medicine held in Stuttgart in July 2004; a revised format at the Wellcome Trust workshop on Medicine and Emotion. It was then read by colleagues, among whom Cathy Kudlik, whose comments I particularly valued and wish to thank here. The research was funded by the Wellcome Trust, the British Academy and the University of Manchester.
2. David Konstan, *Pity Transformed* (London: Duckworth, 2001), p. 12.
3. The history of compassion is older. Tracing the rise of the concept as a 'divinely ordained and sanctioned human expression' found in literature and philosophy from the late seventeenth century onwards, it held a near-unassailable position by the mid-eighteenth century. Norman S. Fiering, 'Irresistible Compassion: An Aspect of Eighteenth-Century Sympathy and Humanitarianism', *Journal of the History of Ideas* (1976), 37, pp. 195–218.
4. Konstan, *Pity Transformed*, pp. 22–3.
5. Charles Darwin, *The Expression of Emotions in Man and Animals* (London: John Murray, 1872).
6. Underlining much of Konstan's work is the work of A. Wierzbicka, e.g. 'L'Amour, la colère, la joie, l'ennui: la sémantique des émotions dans une perspective transculturelle', *Langages* (1988), 89, pp. 97–107. I owe this reference to David Lansglow.
7. While there is much debate as to the exact processes involved in empathy and sympathy, their moral imperative remains endowed with eminent philosophical quality see Martha Nussbaum, *Upheavals of Thought: The Intelligence of Emotions* (Cambridge: Cambridge University Press, 2001).
8. The literature is too large to mention here, but a few titles indicate its dimensions: Sara Ahmed, *The Cultural Politics of Emotion* (Edinburgh: University Press, 2004); David T. Mitchell and Sharon L. Snyder, *Narrative Prosthesis: Disability and Dependencies of Discourse* (Ann Arbor: University of Michigan Press, 2000); Peter Coleridge, *Disability, Liberation and Development* (Oxford: Oxfam, 1996), pp. 3, 47; for a sharp revision of these ideas also see Kimberley Chabot Davis, 'Oprah's Book Club and the Politics of Cross-Racial Empathy', *International Journal of Cultural Studies* (2004), 7, pp. 399–419.

9. Thomas Laqueur, 'Bodies, Details and the Humanitarian Narrative', in Lynn Hunt (ed.), *The New Cultural History* (Berkeley: University of California Press, 1989), pp. 176–204.
10. Laqueur, 'Bodies, Details and the Humanitarian Narrative', p. 176.
11. Mary Poovey, *Making a Social Body* (Chicago: Chicago University Press, 1997), Chapter 2.
12. Jeff D. Bass, 'An efficient Humanitarianism: The British Slave Trade Debates, 1791–1792', *Quarterly Journal of Speech* (1989), 75, pp. 152–65.
13. The debate was notably on the economics of abolitionism. Thomas L. Haskell, 'Capitalism and the Origins of the Humanitarian Sensibility', *The American Historical Review* (1985), 90, pp. 339–61, 547–66; David Brion Davis, John Ashworth, and Thomas L. Haskell in the forum of *The American Historical Review* (1987), 92, pp. 813–28.
14. Karen Halttunen, 'Humanitarianism and the Pornography of Pain in Anglo-American Culture', *The American Historical Review* (1995), 100, pp. 303–34, 324–34.
15. Philippe Solers, 'Réflections solersiennes sur la souffrance', in Samy Abtrann (ed.), *L'Éthique de la souffrance* (Paris: Ellipses, 2000), pp. 60–5.
16. Henry Dunant, *A Memory of Solferino* (London: Cassell and Co., 1947 [1862]), pp. 35–6.
17. For instance Marquis de Sade, *Les Malheurs de Justine* (Paris: Gallimard, La Pléiade, 2000 [1791]).
18. Nicolas Retif de la Bretonne, *Les Nuits révolutionnaires* (Paris: Les Editions de Paris, 1989), pp. 109–14.
19. Retif de la Bretonne, *Les Nuits révolutionnaires,* p. 124.
20. Susan Dunn, 'Michelet and Lamartine: Regicide, Passion and Compassion', *History and Theory* (1989), 28, pp. 275–95; Susan Dunn, *The Deaths of Louis XVI: Regicide and the French Political Imagination* (Princeton, NJ: Princeton University Press, 1994).
21. Carlo Ginzburg, 'Killing a Chinese Mandarin: The Moral Implications of Distance', *Cultural Inquiry* (1994), 21, pp. 46–60, 51; Denis Diderot, Lettre sur les aveugles à l'usage de ceux qui voyent (London [s.n.], 1749), n.p.
22. See Mitchell and Snyder, *Narrative Prostheses,* especially Chapter 3; Martha Stoddard Holmes, *Fictions and Affliction* (Ann Arbor: The University of Michigan Press, 2004), pp. 94–131. Many thanks to Cathy Kudlik for this reference.
23. See Gertrude Himmelfarb, *Poverty and Compassion: The Moral Imagination of the Late Victorians* (New York: Alfred Knopf, 1991), pp. 3–5. Himmelfarb thus makes the point that compassion for the needy took a century to be democratized and extended to the poor at large.
24. David E. Cartwright, 'Kant, Schopenhauer and Nietzsche on the Morality of Pity', *Journal of the History of Ideas* (1984), 45, pp. 83–98, 85–6.
25. Cartwright, 'Kant, Schopenhauer', pp. 88–9.
26. John. E. Atwell, *Schopenhauer on the Character of the World: The Metaphysics of Will* (Berkeley: University of California Press, 1995).
27. Richard J. Behan, *Pain: Its Origin, Conduction, Perception and Diagnostic Significance* (New York and London: D. Appleton and Co., 1915), p. 19.
28. Harriet Martineau, *Life in the Sick-Room* (Peterborough: Broadview, 2003 [1844]), p. 59; Miriam Bailin, *The Sickroom in Victorian Fiction* (Cambridge: Cambridge University Press, 1994), pp. 109–36.

29. Martineau, *Life in the Sick-Room*, pp. 121–4.
30. Honoré de Balzac, 'Les Employés', in *Œuvres complètes* (Paris: Club du Livre, 1965), vol. 5, pp. 1082–83.
31. On Balzac's materialism see David Harvey, *Paris: Capital of Modernity* (London: Routledge, 2003), Chapter 2.
32. Ginzburg, 'Killing a Chinese Mandarin', pp. 49–51.
33. Vautrin is opposed to Père Goriot, the incarnation of virtue here presented as senile and pathetic. Honoré de Balzac, 'Le Père Goriot', in *Œuvres Complètes*, vol. 4, pp. 130–83.
34. Ibid., p.135.
35. The semi-fictional *Mémoires de Vidocq* has played an important role in shaping images of the French police. Rayner Heppenstall, 'Balzac's policemen', *Journal of Contemporary History* (1973), 8, pp.47–56; Jo Burr Margadant, 'Gender, Vice, and the Political Imaginary in Postrevolutionary France: Reinterpreting the Failure of the July Monarchy, 1830–1848', *The American Historical Review* (1999), 104, pp. 1461–96, 1486–7.
36. Dr Conzette, *Note sur certaines formes de la douleur psychique* (Paris: Steinheil, 1913), p. 77.
37. Florence Nightingale, *I Have Done My Duty: Florence Nightingale in the Crimean War, 1854–56*, (ed.) Sue M. Goldie (Manchester: Manchester University Press, 1987); Dunant, *Solferino*.
38. Henry Dunant, Mémorandum au sujet de la société financière et industrielle des moulins de Mons–Djémila, en Algérie au capital de un million, par M. J. Henry Dunant, président de cette société. Exposé des démarches qui ont été faites de 1853 à 1859, et sans interruption, pour obtenir plus de sept hectares de terrain en Algérie (Paris: impr. de Renou et Maulde, 1859).
39. Martin Gumpert, Dunant: *The Story of the Red Cross* (Oxford: Oxford University Press, 1938). For a critical perspective see John F. Hutchinson, *Champions of Charity: War and the Rise of the Red Cross* (Boulder: Westview, 1996).
40. *Conférence internationale pour étudier les moyens de pourvoir à l'insuffisance du Service Sanitaire dans les Armées en Campagne* (Geneva, 26–29 October 1863); *La Charité sur les champs de bataille, Suites du "Souvenir de Solferino"* et résultats de la conférence internationale de Genève (Geneva: n.p., 1864); Geoffrey Best, *War and Law since 1945* (Oxford: Oxford University Press, 1994); idem, *Humanity in Warfare* (London: Methuen, 1980); Michael Howard (ed.), *Restraints on War: Studies in the Limitation of Armed Conflict* (Oxford: Oxford University Press, 1979); G.I. Draper, 'Humanitarianism in the Modern Law of Armed Conflicts', *International Relations* (1985), 8, pp. 380–96; Jacques Meurant, 'Inter Arma Caritas: Evolution and Nature of International Humanitarian Law', *Journal of Peace Research* (1987), 24, pp. 237–49.
41. Dunant, *Solferino*, pp. 35–6.
42. Philippe Ariès, *L'Homme devant la mort* (Paris: Seuil, 1977), p. 465.
43. John Hutchinson, 'Rethinking the Origins of the Red Cross', *Bulletin of the History of Medicine* (1989), 63, pp. 557–78 and idem, *Champions of Charity*; Bertrand Taithe, 'The Red Cross Flag in the Franco–Prussian War: Civilians, Humanitarians and War in the 'Modern' Age', in Roger Cooter, Steve Sturdy and Mark Harrison (eds), *Medicine, War and Modernity* (Stroud: Sutton Publishing, 1998), pp. 22–47.

44. Dunant, *Solferino*, p. 36.
45. On international law see Best, *Humanity in Warfare*, pp. 142–3, which spirited denunciation of Dunant's compromising highlights the war footing of the Red Cross movement.
46. J.C. Chenu, *De la Mortalité dans l'armée et des moyens d'économiser la vie humaine* (Paris: Hachette, 1870), pp. 225–417.
47. See Joshua Cole, *The Power of Large Numbers: Population, Politics and Gender in Nineteenth-Century France* (Ithaca, NY: Cornell University Press, 2000).
48. For a fuller discussion of this debate and its subtleties, see B. Taithe, *Defeated Flesh: Medicine, Welfare, and Warfare in the Making of Modern France* (Manchester: Manchester University Press, 1999), pp. 77–8.
49. L. Chapplain, *De l'Intendance du corps médical et de la mortalité dans l'armée* (Paris: Jean Dumaine, 1872), p. 83.
50. Joris-Karl Huysmans, *Against Nature* (Harmondsworth: Penguin, 1959).
51. Ralph Gibson, 'Hellfire and Damnation in Nineteenth-Century France', *Catholic Historical Review* (1988), 74, 3, pp. 383–402.
52. Richard Burton, *Blood in the City: Violence and Revelation in Paris, 1789–1945* (Ithaca NJ: Cornell University Press, 2001).
53. Ivan Strenski, *Contesting Sacrifice: Religion, Nationalism, and Social Thought in France* (Chicago: University of Chicago Press, 2002); Christopher Forth, *The Dreyfus Affair and the Crisis of French Manhood* (Baltimore: The Johns Hopkins University Press, 2004).
54. Taithe, 'The Red Cross Flag', pp. 22–47,
55. J.M. Charcot, *Les Démoniaques dans l'Art* (Paris: Delahaye, 1887); Josephine Butler, *Catherine of Sienna: A Biography* (London: Dyer, 1885).
56. Ruth Harris, *Lourdes* (London: Allen Lane, 1999), p. 338; Émile Zola, *Lourdes* (Paris: Charpentier, 1898).
57. Charles Martian Allemand-Lavigerie, *Œuvres choisies de son éminence le cardinal Lavigerie* (Paris: Poussielgue Frères, 1884), p. 154, also in Abbé Vincent Burozet, *Histoire des désordres de l'Algérie, 1866–1867–1868, sauterelles, tremblement de terre, choléra, famine* (Algiers and Paris: Garaudel, 1869), p. 78. The latter originated from a report by the Curé of Tenes, Archives Générales des Missionaires d'Afrique, Papiers Lavigerie, B7–240 1.
58. William Ian Miller has reflected on some of the political implications of disgust which apply here. See *The Anatomy of Disgust* (Harvard: Harvard University Press, 1997), pp. 207–10.
59. The concept is evidently informed by Lacanian psychoanalysis. See David Halperin, 'Monsters from the ID', unpublished paper; Didier Eribon, *Hérésie: Essais sur la Théorie de la Sexualité* (Paris: Fayard, 2003); Julia Kristeva, *Powers of Horror: An Essay on Abjection* (New York: Columbia University Press, 1982); Monique Saigal, 'Le Cannibalisme maternel: L'abjection chez Jeanne Hyvrard et Kristeva', *The French Review* (1993), 66, 3, pp. 412–19. On issues of representation see Hal Foster, 'Obscene, Abject, Traumatic', *October* (1996), 78, pp. 106–24.
60. Roselyne Rey, *Histoire de la douleur* (Paris: La Découverte, 1993), pp. 320–42.
61. A similar notion had been observed among survivors of train crashes who attempted to rescue the wounded, most famously Charles Dickens. See Roger Cooter and Bill Luckin (eds), *Accidents in History* (Amsterdam: Clio Medica, Rodopi, 1997); Ralph Harrington, 'The Railway Accident: Train, Trauma and

Technological Crises in Nineteenth-Century Britain' in Mark S. Micale and Paul Lerner (eds), *Traumatic Pasts: History, Psychiatry and Trauma in the Modern Age, 1870–1930* (Cambridge: Cambridge University Press, 2001), pp. 32–55, 51.

62. Dunant, *Solferino*, p. 31.
63. Ibid., p. 44.
64. See the collected writings of Regnaut in Jules Regnaut, La Douleur: sa nature, ses variétés, son expression, son utilité, son utilisation, les moyens de l'éviter et de la combattre (Paris: Médicis, 1947), pp. 12, 168–9, 182–92.
65. Harris, *Lourdes*, pp. 261, 321.
66. The previous religious and medical response had been by the widely discredited Dr Imbert-Goubeyre, *la stigmatisation, l'extase divine et les miracles de lourdes. Réponse aux libres penseurs* (Clermont Ferrand: n.p.,1894), which listed hundreds of cases from the first known case of St Francis.
67. Stanislas Furnet, *Douleur et Stigmatisation* (Paris: Desclée de Brouwer et Cie., Études Carmélitaines, 20ème année, vol II, 1936), 'Introduction', pp.I–II.
68. *Celsus, De Medicina*, Book 7. 4. Many thanks to David Langslow for this reference.
69. Critical analyses have recently flourished to discuss how humanitarians found themselves in a narrative straitjacket of their own writing. See *inter alia*: Barbara Harrell-Bond 'Humanitarianism in a Straitjacket', *African Affairs* (1985), 84, pp. 3–13, 4; John Prendergast, *Crisis Response: Humanitarian Band Aids in Sudan and Somalia* (London: Pluto 1997); also the systematic critique by the late François Jean (ed.) *Life, Death and Aid: the Médecins sans Frontières Report on World Crisis Intervention* (London: Routledge, 1993); Julia Groenewveld, *World in Crisis: Politics of Survival at the end of the Twentieth Century* (London: Routledge, 1997); David Sogge (ed.), *Compassion and Calculation: The Business of Private Foreign Aid* (London: Pluto Press, 1996).
70. David Rieff, *A Bed for the Night: Humanitarianism in Crisis* (New York: Simon and Schuster, 2002); see Alain Woolfolk and Jenny Edkins reviews in *Journal of Human Rights* (2003), 2, 2, pp.247–58.
71. Dunant, *Solferino*, p. 57.

5

Sympathy under the Knife:Experimentation and Emotion in Late Victorian Medicine

Paul White

In his important study of the nineteenth-century controversies over vivisection, Richard French cites this complaint from a review writer in the *Glasgow Medical Journal*:

> In this country, experimental physiology labours under peculiar disadvantages. There still exists some of that feeling of repugnance to vivisections which prevented Sir Charles Bell from discovering those functions of the spinal cord which M. Brown-Sequard is now elucidating. Those tender feelings touching the sufferings of animals, embodied in acts of parliament and incorporated in societies for the prevention of cruelty to animals, not to speak of our innate sympathies, stand in the way of our prosecuting inquiries which demand vivisections for their elucidation.[1]

Though written in 1858, some 15 years before the debate over animal experimentation became widespread in Britain, French offers this passage as a reliable appraisal of the Victorian lay public, and indeed as an indictment of its sentimental and ultimately futile resistance to the advance of medical science. Other commentators on the vivisection debates, and on scientific medicine more generally, have followed suit in their transparent reading of historical sources concerning the inner feelings of medical experimenters and their critics.[2] The persuasiveness of such accounts is perhaps due to the ultimate victory of the 'laboratory revolution in medicine' and to the model of emotions that emerged from the laboratory setting towards the end of the nineteenth century. Recent 'emotional histories', now unfolding on an ambitious scale, indicate the extent to which neurophysiologies and evolutionary psychologies of feeling have flourished sufficiently to provide cultural historians with their objects of study, as it were, ready-made from the laboratory.[3] But how did modern bio-medicine become the authoritative domain in which emotions could be studied, defined, or even transformed? In addressing this question, I want to re-examine the Victorian controversy over vivisection as a debate about the nature of feelings and their place in medical practice.

Remarks such as those by the Glasgow reviewer were highly polemical and widely contested. That researchers had come to master or suppress their 'tender feelings' and 'innate sympathies' was an important theme in accounts of the rise of scientific medicine as a profession, and contributed to the forging of an identity for physiologists as 'men of science'. Indeed, it was the relationship between these two pursuits, the one conceived as a detached search for truth, abstracted from immediate human use, the other as a compassionate dedication to the alleviation of pain and suffering, which would prove most controversial in the debates over vivisection. In the mid-1870s, following a series of graphic reports of experiments on live animals in the mainstream press, public concern over vivisection increased enormously. Legislation was introduced in the Parliament to regulate such experiments, and a Royal Commission was convened to advise on the issue. Medical researchers in this period could not appeal to a scientific authority that was neutral or indisputable. New experimental procedures and diagnostics, anaesthetic technologies and graphic registering instruments, theories of automatism and the evolution of mind and emotion, the recentering of medical education around the laboratory, all of these features of scientific medicine required practical and moral legitimation before a range of audiences, both within and outside the medical profession.

The successful defence of scientific medicine also depended on the ability of experimenters to explain the feelings of their animal subjects, and ultimately to redefine the nature of feeling itself. Faced with the most extensive criticism of experimental practice that Britain had yet seen, experimenters and their supporters presented medical science not merely as a noble advance upon nature, but as triumphant over certain emotions – emotions both within the investigators themselves, who gained critical distance from sentiments which predominated in others, and emotions within the public at large, whose wounded and outraged sympathies sought satisfaction in the regulation of animal experiment. From this debate emerged two opposing images of the laboratory: on the one hand, as a scene of battle, a place of steely nerves and iron discipline, pitted against the forces of disorder, disease and degeneration; and on the other, as a place of senseless slaughter, of sympathies erased and ruthless passions unleashed. The same debate also gave currency to a new model of emotions derived largely from animal experiments, together with a regime of medical diagnostics oriented around laboratory methods and techniques. Thus before 'emotions' became objects of laboratory study as such, the nature and meaning of feelings in the medical sciences was profoundly altered.

Reflexive physiology

Scientific efforts to explain the basis of feeling were crucial to the discipline and display of sentiment in the laboratory and other settings. From the 1850s, physiological writings increasingly depicted feelings as reflex mechanisms, which in persons of high character, were strictly under the control of the will. Among these works perhaps the most prominent in scientific and medical circles were the textbooks of William Carpenter, Fullerian Professor of Physiology at the Royal Institution and later Registrar of the University of London. In the 1856 edition of his *Manual of Physiology*, Carpenter depicted emotions as 'mental feelings' of sensory origin, largely reducible to pleasure or pain. Emotions reacted on the muscular system directly, without the mediation of the intellect or the will. Emotional impulses were closely associated with the sensorimotor activity prevailing among the lower animals. In the higher animals and man, emotions operated in conjunction with intellect, thus forming the springs of action.[4] Carpenter's model depended on a physiological scale of being, in which the powers of body and mind mapped onto a hierarchy culminating in humans. Creatures higher up this scale possessed greater abilities to check the immediate drives of the emotions, to deliberate on their outcomes and, ultimately, to choose between them. This power of self-control, the will, operated in conjunction with the intellect to produce rational and moral behaviour. Activity ensuing purely from emotional impulses was characteristic of creatures lower on the scale, or of pathological conditions, such as the loosely grouped behaviours termed as hysterical. Discipline was achieved through the exercise of abstraction and attention, which directed the mind away from the immediacy of the feelings, so that their strength diminished. Likewise, persistent exposure decreased the intensity of feelings, so that frequent recurrence of the same feeling could result in the adaptation of the system. Thus, for Carpenter, the very highest powers of nature and civilization, as manifest in the intellectual and purposeful capacities of individuals, involved the disciplining of the prime movers of pain and pleasure, the masterful manipulation of emotional forces towards moral ends.

Carpenter's writings were part of a larger literature, which presented what historians of psychology have termed the 'energetic model' of the mind and will.[5] Concerns with channelling or harnessing energy in an efficient and moral manner permeated the writings of Herbert Spencer, Henry Maudsley, Alexander Bain, and Thomas Huxley. The will was the seat of government, presiding over the contest among the lower drives of animal nature, increasingly associated with the higher nervous centres in the cerebral cortex, yet still regarded as irreducible to them.[6] At the same time, from the mid-1850s onwards, this mental physiology was linked to evolutionary accounts of mental life as an improved modification of instinct. Emotions evolved, in all of their fine shades and rich variety, because they facilitated a more subtle and precise adjustment of the organism to its environment.[7] In his 1872 work, *The Expression of the Emotions in Man and Animals*, Darwin explained the full range of emotions as automatic mechanisms, evolved through natural and sexual selection. He also suggested that the science of feelings was itself an expression of the mastery of these powerful mechanisms. In a methodological introduction, Darwin discussed the difficulties inherent in the study of emotions. In fellow humans, such phenomena were often intermixed with each other, or diluted by elements of intellect and volition. When witnessed at first hand, affect could also arouse sympathies in the observer which disrupted the process of investigation. To obtain feelings in their pure form, Darwin chose animals, infants, and natives of Australia, Africa, and India – subjects in whom the emotions were thought to be the least checked by the acquirements of a higher nature and culture. He also introduced research tools such as the questionnaire and photography, which – it was believed – provided distance from his object of study.[8]

Darwin's and Carpenter's works show how the new physiology established an array of hierarchies by means of which the individual mind and body could be ordered as well as the races and creatures of the world. The order operating in these hierarchies relied both on a particular model of feelings as reflex mechanisms, and on a battery of research technologies which brought discipline to potentially disruptive emotions. This physiology was reflexive, for the scientific study of emotions was itself an exercise in the discipline and control of feelings. The power of experimenters to establish the basis of feeling rested on their power, gained through the methods and instruments of science itself, over the mechanical, sensible body.

Feeling with instruments

Among the most important of the new instruments used to measure physiological states were self-registering devices, like the sphygmograph, first developed in 1860 by the French physician Étienne Jules Marey for graphically recording the pulse.[9] Unlike earlier devices such as the mercurial kymograph which required the insertion of a cannula directly into the artery, Marey's instrument was applied to the unbroken skin. A spring-loaded ivory plate rested on the radial artery, transmitting movement to a light wooden recording arm, the tip of which had a knife edge for inscribing the pulse curve on paper or smoked glass. Such an instrument, according to a brief notice in the British medical journal the *Lancet* in 1865, could replace human impressions with 'recorded facts self-analyzed'.[10] Recent literature has examined the development of such graphic technologies as part of the formation of physiology as an authoritative scientific discipline. By converting bodily activities into curves on paper or glass, subject to mathematical analysis, such graphic instruments facilitated the forging of links between physiology and more institutionally established 'hard' sciences, especially physics.[11] Such developments in exact measurement and precision have furnished material for arguments about the rise of mechanical and aperspectival modes of objectivity in the nineteenth century, the systematic suppression of emotional elements from scientific practice, and the cultural separation of the sciences and fine arts.[12] What is perhaps most striking about the new physiological instruments, however, is not that they effaced feelings, but that feelings were displaced, projected from the experimenters onto the apparatus and techniques themselves.

Instruments like Marey's sphygmograph were developed in Germany and France and subsequently introduced into Britain by practitioners, like John Burdon Sanderson, who had trained in overseas laboratories. Before taking up a series of clinical and research posts in London, Burdon Sanderson had studied physiology under Claude Bernard in Paris, following an experimental tradition in which skill and dexterity in the manipulation of instruments was paramount. Burdon Sanderson's own researches – focussed on the influence of respiratory movements on circulation – were initially conducted through sphygmographic and stethographic measurements at Brompton and Middlesex hospitals, where he was a resident physician in the 1860s. He also established a private laboratory in his home. Here, in what appeared to one witness 'an impenetrable jungle of wires', he conducted experiments in electrophysiology, measuring small electrical changes in the active tissues of the heart, and in other muscles of various animals and plants.[13] As professor of Practical Physiology at University College, London, and Superintendent of the new sanitary research establishment, the Brown Institution, Burdon Sanderson was very active in promoting the reorientation of medical education around laboratory sciences such as experimental physiology and histology. Through his lectures and demonstrations, and through a series of practical manuals, he campaigned widely for the introduction of Continental instruments into physiology and medicine.[14]

Perhaps his most important contribution to this effort was the *Handbook of the Physiological Laboratory*, which he produced in conjunction with colleagues at the Brown Institution and University College, Thomas Lauder Brunton and Emmanuel Klein, and with the recently elected Fellow and Praelector of Physiology at Trinity College, Cambridge, Michael Foster. The *Handbook* was a comprehensive manual for the performance of experiments in physiology and histology, written 'for beginners in physiological work'.[15] It was published in two volumes, the second of which was devoted to over 120 plates, including lavish illustrations of the latest physiological instruments. In marked contrast with a comparable Continental publication, Bernard's *Leçons de Physiologie Opératoire,* which showed animals secured to specially constructed tables, or in the process of vivisection, the *Handbook* presented these new technologies utterly divorced from such interventions.[16] With the exception of one plate of the spring myograph, in which an electrode was attached to the leg muscle of a frog, the book deployed an aesthetic in which the animal subjects were erased from the scene of experiment.

Reinforcing this erasure were accompanying descriptions in which the instruments themselves captivated the attention of the experimenter. In his account of an experiment utilizing the sphygmograph, Burdon Sanderson, following conventions laid down by the instrument's inventor, Marey, described the instrument as an organism unto itself. Its functional parts 'communicated' with each other, 'sprang into action', measured and recorded. Accurate performance required the operator to address the instrument with 'great care', 'caution', and 'gentleness'.[17] Similar terms appeared in the sections of the *Handbook* by Foster and Klein, reinforcing the transference of animation to instruments, and of delicacy, sensitivity, and refinement to the relations between instruments and operators. By contrast, the animal subject was addressed only insofar as it was secured to the table. Thereafter, concern was given entirely to the manipulation of registering machines, scalpels, forceps, and hooks, and to the intricacy of operations which revealed the skill and delicacy of the operator.

The sociologist Michael Lynch has interpreted such shifts and erasures in the laboratory setting as a process akin to ritual sacrifice. The experimental procedure, like the religious ceremony, transforms an individual, material body into a universal (or sacred) ideal: a body of knowledge.[18] Graphic registration instruments could transfigure their experimental subjects dramatically, as in Burdon Sanderson's account in the *Handbook* of asphyxia by complete occlusion of the trachea. Here a tube inserted into an artery in the neck communicated with a mercurial manometer and kymograph to record the respiratory convulsions during progressive stages of asphyxiation. A graphic tracing made towards the end of the second minute showed the instant at which convulsions had ceased, and expiration became indistinguishable from inspiration, as death approached and the animal lay 'in a state of tranquillity'.[19] Such experiments not only effaced the sensibility of experimental subjects, but also recombined the sensibility of operators with that of precision machines. At times, the instruments functioned as extensions of the self, but insofar as they were distinct from their operators, they were addressed like animate beings, their sensitivity deserving of respect, their accuracy a reward for attention. In such demonstrations, the laboratory became a social world in which humans and machines interacted in the energetic production of truth.[20]

Brutal passions

Animal experiments were performed in a variety of settings: in private laboratories and research institutes, in lecture halls at university hospitals, and at medical congresses. Public performances, which depended on the manipulation of feelings in the audience as well as the animal subject, were potentially risky. The emotions of witnesses could attach themselves to the experimental subjects, depending on the animal used. Frogs were typically the creatures of choice for experiments performed before non-specialist publics, for they provoked little sympathy.[21] By contrast, domesticated animals such as dogs and cats could elicit the emotional identification of the audience, and transform the scene of experiment dramatically. One notorious demonstration, made in the Smoking Room of the Masonic Hall in Norwich during the meeting of the British Medical Association in 1874, prompted a trial of the demonstrators by the Royal Society for the Prevention of Cruelty to Animals (RSPCA). The operation involved the injection of various alcoholic substances into the thigh of a dog. The experiment was performed in order to observe the animal's convulsions, thus demonstrating the physiological effects of alcoholism, widely regarded in the medical literature of the period as a nervous disease arising from a weakness of the will. Representing the prosecution, the Secretary of the RSPCA, John Colam, testified that the dog had been brought into the room tied to a board and was struggling to free itself from its fetters: 'it was literally crucified, if I may use the term ... struggling and terrified by apprehensions of approaching sufferings'. He also remarked that the presiding operator had entered the room dressed 'not as an ordinary gentleman', but in an apron 'smeared with blood'.[22]

In marked contrast to the image of the laboratory presented in physiology textbooks and hospital schools, as a space of skilful operators and polished instruments, critics of animal experimentation offered graphic descriptions of animals tortured under the knife. In her book *The Modern Rack*, a leading opponent of vivisection, Frances Power Cobbe, utilized plates from Bernard's *Physiologie Opératoire* to display what Burdon Sanderson's *Handbook* had erased, namely, the subjection of animals to science.[23] Cobbe's account operated in a very different register from the works of British and Continental physiologists upon which she drew. It appealed to the very feelings that physiologists claimed worked detrimentally in the experimental setting. Moreover, Cobbe's arguments were underpinned by a quite different account of the nature and significance of the emotional life. She drew upon a tradition of writing in which feeling arose from an affective inner state, by means of a natural process which could not be externally manipulated. This sympathetic tradition had deep roots, for example, in the persona of the 'man of feeling', in whom charitable actions sprang from tender passions, and a heart uncorrupted by vice naturally engendered feelings of benevolence.[24] It also drew upon a counter-evolutionary theory of the sentiments which ranked the spontaneous impulse of sympathy foremost among the mental faculties. Following the moral philosopher William Lecky, Cobbe presented the history of European morals as a progressive development, in which affections were extended first to one's immediate family, then to one's class, to one's nation, to all nations, and finally to the animal world. The bond of sympathy between humans and their domestic animals, particularly dogs, thus represented the pinnacle of moral progress.[25] According to Cobbe, it was the immediacy of this response to suffering that scientific discipline obliterated.

But while Cobbe's critique focussed partly on the apparent heartlessness of the experimental life, it also attacked the unleashing of brutal passions, which were supposed to hold sway among the lower orders of society, the more typical targets of animal protectionists.[26] Thus, the very techniques whereby physiologists demonstrated their mastery and manipulation of feeling could be used by others to drive a moral wedge between the pursuit of science and the humane role that the governing classes were supposed to uphold. Cobbe's criticism of instrumental technologies implied a fundamental questioning of the gentility of scientific character. It would be easier, she remarked, to imagine such operators as ignorant, or hunger-driven, rather than as men at the 'intellectual summit of our social system', the 'calm, cool, deliberate ... the cultivated, the well-fed, the well-dressed, the civilized ... and (it is said) the otherwise kindly disposed and genial men of science'.[27] Initially in favour of regulating vivisection, so that, for example, painful operations would be performed under an anaesthetic, Cobbe later came to promote a total ban on the practice. The 'germ of the disease', she wrote in one of her pamphlets, is 'the *passion itself* which needs to be sternly suppressed; and this can only be done by stopping altogether the practice which is its outcome, and on which it feeds and grows'.

In building their case, anti-vivisectionists compiled a kind of confessional literature from the works of British and Continental physiologists, to display how the pursuit of truth, undiluted by sympathy, could become an all-consuming and unregulated passion. A favourite passage in this genre was culled from a work by the Russian physiologist, Elie de Cyon:

> He who cannot follow some fine nerve-thread, scarcely visible to the naked eye ... with joyful alertness for hours at a time; he who feels no enjoyment when at last, parted from its surroundings and isolated, he can subject that nerve to electrical stimulation ... to such a one there is wanting that which is most necessary for a successful vivisector ... And the sensation of the physiologist, when from a gruesome wound, full of blood and mangled tissue, he draws forth some delicate nervebranch, and calls back to life a function which was already extinguished, – this sensation has much in common with that which inspires a sculptor, when he shapes forth fair living forms from a shapeless mass of marble.[28]

Such passages were designed to work upon their readers in a similar fashion to the graphic sketches of animal cruelty in literature published by the RSPCA, for the equation being made was precisely that between the bestial, anti-social passions of the lower orders, and the intellectual pleasures and excitements of physiological discovery.

Many critics argued that the performance and witnessing of vivisection, increasingly common in medical education by the 1870s, caused practitioners to degenerate, as brutal instincts were heightened through the repeated and prolonged infliction of pain on helpless creatures. George Rolleston, Oxford-educated physician, and Linacre, professor of Anatomy and Physiology at the University, warned against the temptations to which demonstrators were liable as 'the pleasure in experimenting grows stronger by repetition'. He likened vivisections for demonstration purposes to public executions and gladiator shows: 'who does not leave his business and run into the market place so that he may have a look at the drawn sword of the executioner, and ... at the bleeding purple neck of the executed wretch?'.[29] Charles Dodgson (Lewis Carroll) similarly remarked upon the 'thirst for blood' aroused by witnessing carnage, suggesting that in the course of medical education, one's first instincts of horror were gradually deadened by familiarity, turning into a morbid interest, a positive pleasure, and finally 'a ghastly and ferocious delight'.[30]

The sentimental vivisector and the animal automaton

When a Royal Commission convened in 1874 to inquire into the question of vivisection, there was considerable debate about the use of domestic animals. R. H. Hutton, one of the members of the Commission, issued a separate report calling for the exemption of dogs and cats from experiments because they were effectively members of the household. Hutton claimed that in the course of domestication, animals had acquired the same heightened sensitivities that distinguished civilized men from barbarous tribes. It was treacherous and insensible to commit such animals, which were bred and trained up so as to place their confidence in humans, to scientific use. On the other hand, Hutton wrote: 'the tendency of any measure which recognized more explicitly the claims of our family dependants to be especially guarded from anything like hostile treatment, would have a generally humanizing influence on social manners'.[31] In her autobiography, Cobbe professed the same affection for dogs as a country squire: 'Their eagerness ... joyousness [and] transparent little wiles, are to me more winning ... more really and intensely human ... than the artificial, cold and selfish characters one meets too often in the guise of ladies and gentlemen.'[32] Possessed of limited thought, but unlimited love, dogs were creatures whose humble devotion could heal the wounds of disappointed or betrayed affection, and by silent sympathy bring cheer and comfort to the old, poor, or lonely whom the rest of the world abandoned.

A large portion of the English upper and middle classes shared Hutton's and Cobbe's sentiments. As Harriet Ritvo has shown, kindness to animals was not only widely regarded as a mark of general cultivation and moral refinement, but also as a peculiarly English trait.[33] In his *Autobiography*, Darwin described his sensibility towards the suffering of animals, which he attributed to the instruction and example of his sisters.[34] As a Justice of the Peace in the village of Down in Kent, he acted to prevent cruelty to domestic animals, reporting a neighbouring farmer to the RSPCA for working horses with sore necks. He also campaigned with his wife Emma against the use of steel traps on game preserves, co-authoring an appeal that was circulated to landowners and which was based on the premise that 'an English gentleman would not himself give a moment's unnecessary pain to any living creature, and would instinctively exert himself to put an end to any suffering before his eyes'.[35] In his *Life and Letters*, Francis Darwin dwelt at some length on his father's great love for dogs, especially the 'delightfully tender' Polly, and a retriever cross, Bob, whose abrupt change of demeanour from cheerfulness to dejection at the end of his walks was described in *Expression of the Emotions*.[36] Animal experimenters, such as Michael Foster and David Ferrier, often attested to their fondness for dogs and other animals, and this featured in the evidence they presented before the commission on vivisection. Such testimony helped to support claims by other witnesses, such as the physicians James Paget and Joseph Lister, that physiologists on the whole were gentlemanly, with a humane and kind-hearted character.[37]

That vivisectors could be pet owners and animal lovers was proof, it was asserted, that their hearts had not been hardened through laboratory discipline, or by a prolonged exposure to suffering. According to experimenters and their supporters, such warm-heartedness was not evident in the laboratory because only the steadfast control of such sensitive feelings could yield genuine truth.[38] As Samuel Gamgee, a consulting surgeon at Queen's Hospital, asserted in a lecture at the Birmingham Medical Institute in 1882:

> Natural endowments, sense of duty, severity of discipline, may enable a strong man to seem unmoved amidst scenes of the direst woe. But if interrogated, he will often answer in the heroic anguish of Dante's Ugolino,

> 'I wept not: so all stone I felt within.'
>
> Is it not possible to lavish too much sympathy on Magdalenes, murderers, and fools, and to forget the special trials of men and women endowed with aspiring intellects and racking passions, with nerves strung, and hearts thumping, to the utmost of physical endurance?[39]

Like military commanders with countenances steeled for battle at public school, vivisectors underwent a discipline that left them not unsympathetic to the feelings of others, but poised, knife in hand, before grave problems of life and death. This military model, in which the conquest of tender feelings was essential in a struggle against the forces of death, gave a legitimate and triumphalist gloss to the kinds of passions that the critics of vivisection denigrated as selfish and inhumane. Here, the sacrificial subject in the laboratory ritual was not the animal, but the experimenter, who undertook such exhausting and unpleasant labour entirely for the sake of others and at risk to himself.

If a cool exterior was not proof of a lack of sentiment in experimenters, then perhaps the cries and agitations of animals under the knife were not sure indicators of suffering. While practitioners bore frequent witness to the sympathies and compassion of their colleagues, they were far more sceptical of the pains experienced by their experimental subjects. The emotional capital of anti-vivisectionists rested partly on assumptions about the sensitivity of animals, assumptions based in turn on inferences from the outward expressions of animals to their inward feelings. Yet physiological texts by Charles Bell, Marshall Hall, and William Carpenter tended to classify pain among the organism's automatic responses. From the 1850s, evolutionary accounts by Herbert Spencer, Charles Brown-Sequard, and others suggested that the cries and struggles of lower animals were reflex responses, perhaps without accompanying sensations, developed in nature for the protection of creatures against threatened injuries.[40] At the 1874 meeting of the British Association for the Advancement of Science in Belfast, Thomas Huxley pronounced on such arguments in his address, 'On the Hypothesis that Animals are Automata, and its History'.[41] Without declaring absolutely on the questions of animal feeling and the animal soul that had proved controversial since Descartes, Huxley instead emphasized the 'marvellous progress' in the sciences that had ensued on the presumption that such creatures were indeed automata. By denying an identity between an animal's bodily performance and its inner experience, physiologists effectively revised the basis for evaluating feeling. They removed the legitimacy of behavioural analogies between humans and animals, and installed a new definition of feeling that derived from their own research on bodily functions. According to their expert testimony, vivisection was an act in which scientific men could be sympathetic without showing it, while their animal subjects could make a display of feeling without having any. As a member of the Royal Commission on animal experiment, Huxley extracted evidence from Britain's leading physiologists which raised doubt as to whether animals actually experienced pain.[42]

Several of Huxley's colleagues on the Commission, however, pursued a different line of questioning. When George Lewes, who for many years had engaged in physiological researches in his private laboratory, began his testimony by claiming that vivisection was 'a very painful necessity', he was asked by R. H. Hutton which of his own experiments had proved the most painful. Lewes revealed that he had always found the cries of animals distressing, and so had only operated on frogs. The Chairman, Lord Cardwell, Vice President of the RSPCA, then interrupted a protracted discussion on the locus of sensibility and suffering by asking Lewes whether 'in the training of medical students in physiology it is necessary that they should see experiments performed before their eyes?'[43] Hutton and Lord Cardwell, both animal protectionists, introduced a discourse of sympathy to link questions which Huxley and his scientific colleagues were trying to prise apart: namely, the question of animal feeling and the question of human feeling for animals. For the critics of vivisection, debates over the pain experienced by experimental subjects were joined with concerns about the emotions appropriate in the experimenters and witnesses. To respond to such critics, it was insufficient to discount the performances of animals under the knife. The defenders of vivisection needed to mount their own displays of sympathy in the laboratory setting. Such displays deployed another technology for the suppression and manipulation of feeling: anaesthetics.

At the time of the Royal Commission, chloroform and ether had only recently begun to enter common usage among medical practitioners. From the middle of the century onwards, however, promoters of anaesthesia had portrayed the new technology as both liberating and desirable for the patient. It was claimed that anaesthetics facilitated the radical interventions that had transformed surgery from the last resort of a second-class healer to a treatment of choice practised by new cultural heroes.[44] In the *Lancet* in 1872, F. J. Grant called upon his medical peers to 'witness the calm composure of the surgeon and placid sleep of the patient' in the new anaesthetized operations.[45]

Medical authority could now derive from the ability, hitherto denied the operator in the spectacular and agonizing procedures of the past, to remove pain from the subject of the operation. Like the new surgical operations under ether and chloroform, animal experiments previously regarded as brutal could now be considered humane.

Many of the criticisms directed against Continental physiologists in the years leading up to the Royal Commission had focussed on the failure of experimenters to use anaesthetics. In an 1873 review of Burdon Sanderson's *Handbook*, a writer for *Nature* similarly complained that the authors did not specify the use of anaesthetics in any of the operations they described.[46] Such criticism did not go unheeded. In the introduction to his own 1876 textbook, Michael Foster announced that all the experiments set down in the manual could be rendered painless.[47] Likewise, each of the contributors to the *Handbook* affirmed before the Royal Commission their commitment to minimizing or eliminating animal suffering through the administration of new anaesthetic technologies. Such testimonies, however, glossed over disputes about the actual effects of such substances and the motives for their application.

Experts disagreed over whether the use of anaesthetics so altered an animal's physiology as to compromise the results of an experiment.[48] But as George Lewes attested: 'If you have an anaesthetized animal it lies like a log of wood before you ... you can make the most delicate operation with certainty'.[49] Histories of physiology have tended to credit the introduction of anaesthetics not to concerns for pain or suffering, but to considerations of experimental control.[50] Some historians of medicine have also shown how the use of anaesthetics conferred to medical practitioners an unprecedented control over their patients.[51] In a letter to the *Morning Post* in 1875, George Hoggan, a recent medical graduate at Edinburgh, denounced the use of anaesthetics in experiments that he had observed while on a four-month course in the laboratory of Claude Bernard. Experimenters, Hoggan wrote, 'are always ready to repudiate any implied want of tender feelings, but I must say that they seldom show much pity; on the contrary, in practice they frequently show the reverse'. Anaesthetics were 'the greatest curse' to animals, and 'far more efficacious in lulling public feeling towards vivisectors than pain in the vivisected'.[52]

To discount such witnesses, it was crucial for experimenters to establish that their use of anaesthetics stemmed from humane motives, that is, from the same feelings of pity and remorse that flowed spontaneously from the sympathetic public. Perhaps the key testimony in this regard came from Emmanuel Klein, Lecturer in Histology at St Bartholomew's Hospital, one of the contributors to Burdon Sanderson's

Handbook. When asked by the Royal Commissioners whether he used anaesthetics for his own researches, Klein expressed his frank doubts as to the validity of experiments performed under such conditions. When questioned about vivisections performed for demonstration purposes, he was unequivocal. He did use anaesthetics. This was not on account of the animal's pain, however, but for 'convenience's sake':

> Q: ... do you mean that you have no regard at all to the sufferings of the animals?
> K: No regard at all.
> Q: You are prepared to establish this as a principle which you approve?
> K: I think that with regard to an experimenter, a man who conducts special research, and performs an experiment, he has no time, so to speak, for thinking what will the animal feel or suffer. His only purpose is to perform the experiment, and to perform it as quickly as possible.
> Q: Then for your own purposes, you disregard the question of the suffering of the animal in performing a painful experiment?
> K: I do.
> Q: Why do you regard it, then, when it is for demonstration?
> K: Because I know that there is a great deal of feeling against it in this country, and when it is not necessary, one should not perhaps act against the opinion or belief of certain individuals of the auditorium. One must take regard of the feelings and opinions of those peoples before whom one does the experiments.[53]

Anaesthetics were in effect a powerful means by which Klein sought to control his audience as well as his experimental subjects. By calling attention to his own artifice, however, Klein spoiled the show. His remarks were notorious for undoing the works of other witnesses, as well as the elaborate moral and theoretical case advanced by Huxley. His attempts to alter his words in the published report of the Commission failed, and he was cited repeatedly in a wide range of pamphlet literature as evidence that the worst practices of Continental physiology now occupied a central place in British institutions of research and medical education.[54]

The spinster with the lapdog or the medical Tartuffe?

The anti-vivisection campaign has often been regarded as a women's movement, notwithstanding the fact that it was supported by clergymen, statesmen, and men of letters, as well as writers, artists, and philanthropists of both sexes. Animal protection campaigning was indeed, however, one of the few activities in which women could play a public role. That women like Cobbe, who also advocated women's suffrage, could occupy positions of leadership was unusual in the period. But the persistent identification of the cause with women is also an artefact of the gendering of emotion within medical science. It constitutes a legacy of these debates which ensures that women continue to be regarded as reservoirs of sentimental feeling, and therefore as liable to clouded judgment, a source of ignorance, and ineptitude in scientific matters. The successful defence of the experimental life against public sentiment and state legislation was accomplished in part by invalidating, on emotional grounds, the criticism of those who were not themselves medical researchers. The same etiology that translated feminine virtues such as sympathy into female maladies such as 'emotional instability' also served as an argument for excluding women from medical education. The authority of women as social commentators was highly circumscribed, and ultimately failed to diminish the status of those who were accredited as both experts and gentlemen, and thus capable of regulating their own feelings. The horrors of vivisection were packaged for public consumption in popular novels of the period, as in the form of Dr Benjulia in Wilkie Collins' *Heart and Science*. Such characters were monstrous foreigners, who preyed upon innocent women and other weaker creatures in dark sequestered basements; they were not the respectable consultants or professors of English hospitals and universities.

The gendering of scientific and medical expertise had been present in reflex physiology from its origins. It was expressed in Carpenter's and Darwin's works, for example, through hierarchies which ranked the intellect above the emotions, and accorded the sexes different degrees of rational and volitional power.[55] By gendering the sympathetic tradition in which most philanthropists operated, experimenters could also denigrate their male critics, like R. H. Hutton, Huxley's colleague on the Royal Commission, whom Darwin described as resembling 'a *female* Miss Cobbe'.[56] Such gender critiques were thus double-edged, for if the men who opposed vivisection were unmanly, the women who did so could be portrayed as unwomanly. For women to combat the claims of professional men in the public sphere was a distortion of their feminine nature. Prominent women like Cobbe were accused of looking to animals to supply the love they had failed to awaken in men. The figure of the heroic medical professional, toiling at the physiological frontier, was juxtaposed with that of the spinster with the lapdog, hurling hysterical resentment from within an impoverished domestic life. As one commentator wrote: 'Let my adversaries contradict me, if they can show among the leaders of the agitation one young girl, rich, beautiful and beloved, or one young wife who has found in her home the full satisfaction of her affections.'[57]

In this climate of exchange, some anti-vivisectionists counselled against appeals to sentiment. The philanthropist and travel writer Isabel Burton described for her readers how, faced with the prospect of committing a sick animal to authorities who might experiment on it, she resolved to kill her own terrier Nip with six ounces of pure chloroform.[58] Others pressed the issue of sexual politics still further, rather than adopt the cool countenance of their scientific opponents. In an anonymous pamphlet entitled 'The Woman and the Age', the scientific materialism sustaining the hypothesis that animals were automata was converted into a rape of (female) Nature, whose law was sympathy, and who only disclosed her secrets to those 'who court her with reverence, humility, patience, and tenderness'.[59] Still harsher critics associated avidity for truth with the bloodlust of surgeons and the prurience of medical students, explicitly linking women, as objects of the new science of gynaecology, with the victims of animal experiment.[60]

Given the circumlocutions of Huxley and others concerning the emotional life of animals, and the revelations of Klein and certain Continental physiologists on the inner feelings of experimenters, it is perhaps not surprising that the proponents of medical science were often charged with duplicity. By mid-century, there was already nostalgia within the medical profession for the family doctor, the hero of many popular novels of the period, who was at once a sympathetic healer and a family friend and confidant. Irvine Loudon's claim that the moral authority of practitioners actually declined with the rise of medical science is borne out by a range of criticisms that severed the roles of the compassionate physician and the disciplined experimenter.[61] British practitioners, remarked Dodgson, lacked the candour and courage to admit, as did their colleagues on the Continent, that their true motives were selfish, namely, the pursuit of knowledge divorced from considerations of human or animal welfare. As the drive of personal ambition replaced that of philanthropic concern, another critic asserted, medicine had become a 'parvenue profession':

> It is the misfortune of the Medical profession that the performance of its ordinary duties involves the appearance of humane feelings, which may or may not be present on any particular occasion, but

> which the patient and his friends usually expect to see exhibited, and which the doctor is consequently almost driven to simulate ... when the doctor has adopted his profession as a mere gagne-pain, or from taste for science rather than love of humanity, then a certain affectation of sympathy with his patients and their afflicted friends is forced upon him, and we behold the not very rare phenomenon of the medical Tartuffe ... a display of sympathy is part of the stock-in-trade of a physician (especially of one who attends ladies), without which he could not hope for a large clientele, any more than a grocer would succeed in business who failed in civility to his customers.[62]

Many within the medical profession shared misgivings about passionate dedication to the experimental life. Henry Wentworth Ackland, a Regius Professor of Medicine at the University of Oxford, expressed his reservations about the new pursuit of 'biological knowledge' by men whose 'profession is to discover', rather than by 'medical men ... for the purpose of doing immediate good to mankind'.[63] As the work by Gerald Geison and Christopher Lawrence has shown, resistance to the new laboratory disciplines was strong particularly among elite physicians, whose own authority derived from a classical education and clinical experience.[64] The graphic registering instruments promoted by Burdon Sanderson and others were a direct challenge to the tacit learning which gentlemanly practitioners demonstrated by feeling the pulse. Concerns about the use of such technologies for diagnostic operations were also part of more widespread anxieties about the growth of large metropolitan clinics, with their adjacent laboratories and medical schools, where patients might be treated like scientific subjects.[65]

Thus at stake in the controversies over vivisection was not only the moral character of experimental physiologists, but also the public image of medical practitioners, as their profession underwent extensive reforms. These included the introduction of laboratory practicals in physiology, the creation of new lectureships and chairs in physiology at Oxford and Cambridge, and the erection of new laboratories on a large scale. By the turn of the century, the physiological laboratories at University College, London, St Bartholomew's Hospital, and Cambridge University were equipped with the latest experimental and diagnostic technologies, supplied by a burgeoning industry of instrument makers. The Cambridge laboratory was fitted with 16 drum kymographs, driven by a pulley mechanism, and each linked to a battery, electric key, and spring myograph, arranged across a long table, thus enabling students to work simultaneously.[66]

Conclusion: The medicalization of emotion

The vivisection controversy of the 1870s has often been studied as part of the history of scientific and medical professionalization, involving the development of graphic recording instruments and highly impersonal styles of communication, the establishment of the laboratory as an objective space, and the rise of an autonomous community of experts with the authority to pursue their inquiries without scrutiny from the sentimental public. I have interpreted these developments not as documenting the effacement of emotions from science, but rather as raising new kinds of questions about the nature of feelings and their place in a wide range of settings. Underpinning the authority of the new laboratory-based life sciences and medicine were displays of passion for animals, for science, and for humanity. These displays, made in a variety of realms, hospitals, lecture theatres, print, and arguably, the laboratory itself, exhibited the ability of practitioners to exert masterful control over the movement of their feelings, and to command authority in a wide range of domains. Next to the image of the medical researcher as a detached or mechanical observer, we must place the highly gendered, impassioned discoverer and sympathetic investigator, concerned for the well-being of his subjects through the administration of anaesthetics. The control of emotion within the experimental setting could itself be described as a highly affective process: vivisection required 'aspiring intellects and racking passions, with nerves strung, and hearts thumping, to the utmost of physical endurance'. Together with the neutral discourse of textbooks and monographs, we find a mode of scientific reporting in which instruments became animated, and their relations with practitioners affective.

Such discourse and displays were indeed challenged by accounts of experimenters as both unfeeling, that is, unsympathetic towards their animal subjects or human fellows, and also undisciplined, at the mercy of passions for truth and selfish gain which created a pleasure in causing suffering. Critics described the arrogance of the new medical 'men of science', and the inhumanity of institutions in which patients were treated as clinical material. In an address to the London Anti-Vivisection Society in 1890, George Bernard Shaw remarked: 'Who cares about the bile ducts in comparison with the human affections and passions? Why not take men and women and, having put sphygmographs on their wrists, thermometers under their tongues, and every other appliance at your command for measuring the effect of emotions on the body, proceed to play upon their feelings.'[67] One book, *The Nine Circles of Hell*, described a series of animal experiments of spiralling intensity, from poisoning with alcohol and absinthe, to suffocation by apnoea and drowning, burning and freezing, and finally, 'moral experiments' on 'the feelings of dogs'. One example of the latter described a physician who 'inspired a dog with the greatest aversion for [him] by plaguing and inflicting some pain or other upon it, as often as [he] saw it'.[68]

Perhaps no English 'gentleman of science' would have experimented on dogs in quite this way. Darwin's research on emotional expression relied largely on 'field' observations and anecdotal evidence. Yet it also displayed, particularly in its use of photography, an interest in technologies that could measure and even provoke feeling in new ways. Darwin made considerable use of images created in the laboratory of the physiologist Guillaume Benjamin Duchenne, depicting a wide range of emotional expressions induced not by inner feeling, but by electrical stimulation of the face. In the course of his research, Darwin also corresponded extensively with the alienist James Crichton Browne, and received dozens of photographs of patients taken in the West Riding Asylum where Browne was the director.[69] Bearing the physician's own handwritten diagnoses, these photographs show how emotional conditions and character were inscribed with a typology of mental illness, and thereby operated within the larger experimental regime at the Asylum in conjunction with studies of brain localization and cerebral function. Towards the end of the nineteenth century emotions had not yet entered the laboratory as objects of study *per se*.[70] Yet many of the conditions necessary for such clinical researches of the early twentieth century had been satisfied, namely the transformation of emotions into material, measurable entities, along with the instrumentation and, it might be added, the moral licence, necessary to record and analyse them.

Notes

1. Anonymous, 'The Salivary Glands and Pancreas, Their Physiological Actions and Uses in Digestion: Being a Review of the Doctrines Taught by M. Claude Bernard', *Glasgow Medical Journal* (1858), xii, pp. 1–2; cited in R. French, *Antivivisection and Medical Science in Victorian Society* (Princeton: Princeton University Press, 1975), p. 38.
2. Harriet Ritvo, *The Animal Estate: The English and Other Creatures in the Victorian Age* (Cambridge, Mass.: Harvard University Press, 1987), pp. 157–66; James Turner, *Reckoning with the Beast: Animals, Pain, and Humanity in the Victorian Mind* (Baltimore: Johns Hopkins University Press, 1980); Nicholaas A. Rupke (ed.), *Vivisection in Historical Perspective* (London: Croom Helm, 1987); Stewart Richards, 'Drawing the Life-Blood of Physiology: Vivisection and the

Physiologists' Dilemma, 1870–1900,' *Annals of Science* (1986), 43, pp. 27–56; William F. Bynum, *Science and the Practice of Medicine in the Nineteenth Century* (Cambridge: Cambridge University Press, 1994), p. 171.
3. For example, William M. Reddy, *The Navigation of Feeling: A Framework for the History of Emotions* (Cambridge: Cambridge University Press, 2001); and Peter N. Stearns and Jan Lewis (eds), *An Emotional History of the United States* (New York: New York University Press, 1998).
4. William B. Carpenter, *A Manual of Physiology*, 3rd ed. (London: Churchill, 1856), p. 596.
5. Kurt Danziger, *Naming the Mind: How Psychology Found its Language* (London: Sage Publications, 1997), pp. 51–65.
6. Janet Oppenheim, *'Shattered Nerves': Doctors, Patients, and Depression in Victorian England* (New York: Oxford University Press, 1991), pp. 141–51.
7. For example, Herbert Spencer, 'On Laughter', in *Essays, Scientific, Political, and Speculative*, 2nd Series (London: Williams and Norgate, 1863), p. 111.
8. Charles Darwin, *The Expression of the Emotions in Man and Animals* (London: Murray, 1872), pp. 13–19.
9. See Robert Frank Jr., 'The Tell-Tale Heart: Physiological Instruments, Graphic Methods; and Clinical Hopes, 1854–1914', in William Coleman and Frederic L. Holmes (eds), *The Investigative Enterprise: Experimental Physiology in Nineteenth-Century Medicine* (Berkeley: University of California Press, 1988), pp. 211–90; and Christopher Lawrence, 'Physiological Apparatus in the Wellcome Museum: 1. The Marey Sphygmograph', *Medical History* (1978), 22, pp. 196–200.
10. Anonymous, 'Physicians and Physicists', *Lancet* (1865), 2, p. 599.
11. Soraya de Chadarevian, 'Graphical Method and Discipline: Self-Recording Instruments in Nineteenth-Century Physiology', *Studies in History and Philosophy of Science* (1993), 24, pp. 267–91; Frederic L. Holmes and Kathryn M. Olesko, 'The Images of Precision: Helmholtz and the Graphic Method in Physiology', in M. Norton Wise (ed.), *The Values of Precision* (Princeton: Princeton University Press, 1995), pp. 198–221; Robert Michael Brain, 'The Graphic Method: Inscription, Visualization and Measurement in Nineteenth-Century Science and Culture', Unpublished PhD, University of California, Los Angeles, 1996.
12. Lorraine Daston, 'Objectivity and the Escape from Perspective', *Configurations* (1992), 22, pp. 597–618, and Lorraine Daston and P. Galison, 'The Image of Objectivity', *Representations* (1992), 40, pp. 81–128; George Levine, *Dying to Know: Scientific Epistemology and Narrative in Victorian England* (Chicago: University of Chicago Press, 2002).
13. Walter John O'Connor, *Founders of British Physiology*, 1820–1885: *A Biographical Dictionary* (Manchester: Manchester University Press, 1988), p. 143. See also Terrie M. Romano, *Making Medicine Scientific: John Burdon Sanderson and the Culture of Victorian Science* (Baltimore: Johns Hopkins University Press, 2002).
14. John Scott Burdon Sanderson, *Handbook of the Sphygmograph* (London: Robert Hardwicke, 1867).
15. Idem (ed.), *Handbook for the Physiological Laboratory*, 2 vols (London: J. & A. Churchill, 1873), p. vii.
16. Claude Bernard, *Leçons de Physiologie Opératoire* (Paris: J. B. Baillière et fils, 1879).

17. The use of affective language in the *Handbook* is noted in Stewart Richards, 'Vicarious Suffering, Necessary Pain: Physiological Method in Late Nineteenth-Century Britain', in Rupke (ed.), *Vivisection*, pp. 125–48.
18. Michael Lynch, 'Sacrifice and the Transformation of the Animal Body into a Scientific Object: Laboratory Culture and Ritual Practice in the Neurosciences', *Social Studies of Science* (1988), 18, pp. 265–89. On aesthetic, performative, and emotional aspects of laboratory life, see Nicholas Jardine, 'The Laboratory Revolution in Medicine as Rhetorical and Aesthetic Accomplishment', in Andrew Cunningham and Perry Williams (eds.), *The Laboratory Revolution in Medicine* (Cambridge: Cambridge University Press, 1992), pp. 304–23.
19. Burdon Sanderson (ed.), *Handbook*, pp. 319–20.
20. On the sociable relations between instruments and operators in the laboratory setting, see Karin Knorr-Cetina, 'Sociality with Objects: Social Relations in Post-Social Knowledge Societies', *Theory, Culture and Society* (1997), 14, pp. 1–30, and eadem, *Epistemic Cultures: How the Sciences Make Knowledge* (Cambridge, MA: Harvard University Press, 1999).
21. Frederic Holmes, 'The Old Martyr of Science: The Frog in Experimental Physiology', *Journal of the History of Biology* (1993), 26, pp. 311–28.
22. *Report of a Prosecution of Physiologists by the R. S. P.C. A., at the Town Hall, Norwich, for Alleged Cruelty to Two Dogs* (London: Robert Hardwicke, 1875), pp. 5–8.
23. Francis Power Cobbe, *The Modern Rack. Papers on Vivisection* (London: Swan and Sonnenschein and Co., 1889).
24. Ronald S. Crane, 'Suggestions Toward a Genealogy of the "Man of Feeling", *English Literary History* (1933), 1, pp. 205–30.
25. Francis Power Cobbe, *Darwinism in Morals, and Other Essays* (London: Williams and Norgate, 1872). See also the discussion of animal sympathy as a recent development in the progress of moral sentiments in William Lecky, *History of European Morals* (London: Longmans, Green, and Co. 1865), pp. 182–8.
26. Ritvo, *The Animal Estate*, pp. 144–8; B. Harrison, 'Animals and the State in Nineteenth-Century England', *English Historical Review* (1973), 87, pp. 786–820.
27. Francis Power Cobbe, *Life of Frances Power Cobbe by Herself*, 2 vols (London: Richard Bentley & Son, 1894), vol. 2, p. 290.
28. Elias de Cyon [Ilya Faddeyich Tsion], *Methodik der Physiologischer Experimente und Vivisectiones* (Geissen and St. Petersberg: Ricker, 1876), p. 15. Cited in Cobbe, *The Modern Rack*, pp. 197–8.
29. *Report of the Royal Commission on the Practice of Subjecting Live Animals to Experiments for Scientific Purposes; with Minutes of Evidence and Appendix. Presented to both Houses of Parliament by Command of her Majesty* (London: George Eyre and William Spottiswoode, 1876). The testimony of Rolleston is given in Q. 1729–30, 4908–09.
30. Lewis Carroll, *Some Popular Fallacies about Vivisection* (Oxford: Privately printed, 1875), pp. 8–9.
31. *Report of the Royal Commission, pp. xxii–xxiii.*
32. Cobbe, *Life*, vol. 2, pp. 241–2.
33. Ritvo, *The Animal Estate*, pp. 126–35.
34. N. Barlow (ed.), *The Autobiography of Charles Darwin* (London: Collins, 1958), pp. 26–7.

35. The appeal, published privately in 1864, is reproduced in Frederick Burkhardt and Sydney Smith, (eds), *The Correspondence of Charles Darwin* (Cambridge: Cambridge University Press 1999), vol. 12, Appendix IX.
36. Francis Darwin (ed.), *Life and Letters of Charles Darwin*, 3 vols (London: Murray, 1888), vol. 1, pp. 113–15; Darwin, *Expression of the Emotions*, pp. 57–8.
37. *Report of the Royal Commission*. See Q. 2346 (Foster), Q. 3249–50 (Ferrier), Q. 365 (Paget), Q. 4368–72 (Lister).
38. For example, Ludimar Hermann, *The Vivisection Question Popularly Discussed* (London: Williams and Norgate, 1877), pp. 1–4; and *A Speech Delivered by Professor Humphrey, F. R. S., at the 49th Annual Meeting, Ryde, August, 1881* (London: British Medical Association, 1881).
39. Sampson Gamgee, *The Influence of Vivisection on Human Surgery*, 2nd ed. (London: J. and A. Churchill, 1882), pp. 27–8.
40. Roselyne Rey, *The History of Pain* (Cambridge: Cambridge University Press, 1995), chapter 6.
41. Thomas H. Huxley, 'On the Hypothesis that Animals are Automata, and its History', *Nature* (1874), 10, pp. 362–6. Republished in *Collected Essays*, 9 vols (London: Macmillan, 1894), vol. 1, pp. 199–250.
42. *Report of the Royal Commission*. See Q. 2371–9 (Burdon Sanderson), Q. 4375–92 (Lister), and Q. 3406–11 (Ferrier).
43. *Report of the Royal Commission*. See Q. 6411–2.
44. Christopher Lawrence, 'Democratic, Divine, and Heroic: The History and Historiography of Surgery', in idem (ed.), *Medical Theory, Surgical Practice* (London: Routledge, 1992), pp. 1–47. See also Martin S. Pernick, *A Calculus of Suffering: Pain, Professionalism and Anaesthesia in Nineteenth Century America* (New York: Columbia University Press, 1985); Peter Fairley, *The Conquest of Pain* (London: Michael Joseph, 1978); and Jürgen Thorwald, *The Triumph of Surgery* (London: Thames and Hudson, 1960).
45. *Lancet* (1872), p. 404.
46. *Nature* (1873), p. 441.
47. Michael Foster and John Newport Langley, *A Course of Elementary Practical Physiology* (London: Macmillan, 1876), p. viii.
48. See, for example, [G. Jesse], *To the People of Great Britain and Ireland. Correspondence with Thomas Lauder Brunton, M.D., F. R. S., D.Sc.,* (from 18th February 1876, till 11th April, 1879.) *on Nitrite of Amyl in Angina Pectoris* (London: Society for the Total Abolition and Utter Suppression of Vivisection, 1879).
49. *Report of the Royal Commission*, Q. 6376.
50. For example, Karl Rothschuh, *History of Physiology* (New York: Krieger, 1973).
51. Alison Winter, 'Ethereal Epidemic: Mesmerism and the Introduction of Inhalation Anaesthesia in Early Victorian London', *Social History of Medicine* (1991), 4, pp. 1–27.
52. *Morning Post*, 2 February 1875. Reprinted in *Spectator* (1875), 48, pp. 177–8.
53. *Report of the Royal Commission*, Q. 3539–42.
54. For example, Mona Caird, *A Sentimental View of Vivisection* (London: William Reeves, 1894).
55. On the subordination of feeling in the hierarchy of the senses, and its associations with feminine weakness, see Sander Gilman, 'Touch, Sexuality and

Disease', in William F. Bynum and Roy Porter (eds), *Medicine and the Five Senses* (Cambridge: Cambridge University Press, 1993), pp. 198–224; and Cynthia Russett, *Sexual Science: The Victorian Construction of Womanhood* (Cambridge, MA: Harvard University Press, 1989).

56. Darwin's remark is reported in the *British Medical Journal* (1904), 1, p. 1266.
57. Elias de Cyon, 'The Anti-Vivisectionist Agitation', *Contemporary Review* (1883), 43, p. 506; Francis Power Cobbe, 'Hysterical Old Maids', *Englishwoman's Review*, 15 June 1883, pp. 249–50. See also Mary Ann Elston, 'Women and Anti-vivisection in Victorian England, 1870–1900' in Rupke (ed.), *Vivisection*, pp. 260–82; and Hilda Kean, 'The "Smooth Cool Men of Science": The Feminist and Socialist Response to Vivisection', *History Workshop Journal* (1995), 40, pp. 16–38.
58. Isabel Burton, *Prevention of Cruelty, and Anti-Vivisection* (London: William Mullan and Son, 1879), p. 4.
59. *"The Woman" and The Age: A Letter Addressed to the Right Hon. W. E. Gladstone, M.P.* (London; E. W. Allen, 1881).
60. See Ann Dally, *Women under the Knife: A History of Surgery* (London: Hutchinson Radius, 1991); and Coral Lansbury, 'Gynaecology, Pornography and the Anti-vivisection Movement', *Victorian Studies* (1985), 28, pp. 413–37.
61. Irvine S. L. Loudon, 'The Concept of the Family Doctor', *Bulletin of the History of Medicine* (1984), 58, pp. 347–62.
62. Anonymous, *The Medical Profession and Its Morality* (London: Pewtress and Co., 1881), pp. 14–15.
63. *Report of the Royal Commission*, Q. 944.
64. Christopher Lawrence, 'Incommunicable Knowledge: Science, Technology and the Clinical Art in Britain, 1850–1914', *Journal of Contemporary History* (1985), 20, pp. 503–20; Gerald L. Geison, *Michael Foster and the Cambridge School of Physiology: The Scientific Enterprise in Victorian Society* (Princeton: Princeton University Press, 1978), pp. 33–47. See also Samuel Edward Doyle Shortt, 'Physicians, Science and Status: Issues in the Professionalization of AngloAmerican Medicine in the Nineteenth Century', *Medical History* (1983), 27, pp. 51–68.
65. N. D. Jewson, 'The Disappearance of the Sick Man from Medical Cosmology, 1770–1870', *Sociology* (1978), 10, pp. 225–44; Susan E. Lederer, *Subjected to Science: Human Experimentation in America before the Second World War* (Baltimore: Johns Hopkins University Press, 1995).
66. On these transformations in medical education, see Merriley Borell, 'Training the Senses, Training the Mind', in Bynum and Porter (eds), *Medicine and the Five Senses*, pp. 244–61.
67. George Bernard Shaw, *The Dynamitards of Science* (London: The London Anti-Vivisection Society, 1900), p. 18.
68. G. M. Rhodes, *The Nine Circles of the Hell of the Innocent. Described from the Reports of the Presiding Spirits* (London: Swan Sonnenschein and Co., 1892), p. 157.
69. G.-B. Duchenne de Boulogne, *Mécanisme de la physionomie humaine ou analyse électro-physiologique de l'expression des passions applicable à la pratique des arts plastiques* (Paris: Jules Renouard, 1862). On Darwin's collection of photographs and his interest in the work of Duchenne and Crichton Brown, see Janet Browne, 'Darwin and the Face of Madness', in William F. Bynum,

Roy Porter, and Michael Shepherd (eds), *The Anatomy of Madness. Essays in the History of Psychiatry. Vol. 1, People and Ideas* (London: Tavistock Publications, 1985), pp. 151–65; Phillip Prodger, 'Illustration as Strategy in Charles Darwin's "*The Expression of the Emotions in Man and Animals*"', in Timothy Lenoir (ed.), *Inscribing Science: Scientific Texts and the Materiality of Communication* (Stanford: Stanford University Press, 1988), pp. 140–81; and Paul Ekman (ed.), 'Photography and the "Expression of the Emotions"', in Charles Darwin, *The Expression of the Emotions in Man and Animals* (London: Harper Collins, 1988), pp. 399–415.

70. On the clinical study of emotions in the twentieth century, see Otniel E. Dror, 'Techniques of the Brain and the Paradox of Emotions, 1880–1930', *Science in Context,* (2001), 14, pp. 643–60; idem, 'The Affect of Experiment: The Turn to Emotions in Anglo-American Physiology, 1900–1940', *Isis* (1999), 90, pp. 205–37; and idem, 'The Scientific Image of Emotion: Experience and Technologies of Inscription', *Configurations* (1999), 7, pp. 355–401.

6
Fear and Loathing in the Laboratory and Clinic

Otniel E. Dror

While Professor Pagliani was standing before the registering apparatus, with his arms in the glass cylinders filled with water, Professor Ludwig walked into the room. Immediately the two pens indicating the volume of the arms, descended, as though a vertical line, ten centimetres in length, were drawn down this page. It was the first time that I had seen such a considerable decrease in the volume of the hand and forearm, produced by an apparently slight emotion. Professor Ludwig himself was very much astonished, and, with that affability which made him so beloved by his pupils, took a pen and wrote on the paper at that point where the plethysmograph had marked the disturbance in the circulation caused by his appearance, *Der Löwe Kommt* (Enter the lion).[1] Angelo Mosso, *Fear*, 1896

It is only recently that 'the marvelous minuteness of ... [the] response of the body to the soul has been established by the experiments of prominent psychologists and physiologists.[2] *The Literary Digest*, 1914

We are constantly developing more accurate methods of measuring human emotions... .It is possible, then, to speak of the emotion quantitatively, as being present in large or small amounts.[3] *The New York Times*, 1925

In his well-known study *The Human Motor*, Anson Rabinbach argued for the centrality of fatigue in late-nineteenth- and early-twentieth-century European societies. Fatigue, which Rabinbach characterized as the 'permanent nemesis of an industrializing Europe,' acquired its modern significance in conjunction with the emergence of European concerns with energy, work, and efficiency. The major late-nineteenth-century figure who, according to Rabinbach, 'placed the study of fatigue within the canon of mechanical materialism' and 'invented the first efficient and accurate measure of fatigue' – the ergograph – was the Italian physiologist Angelo Mosso.[4] Although Rabinbach's study begins with Mosso's work on fatigue, Mosso came to the study of fatigue only after a decade of intensive laboratory investigations of mental and affective states. His materialistic approach to fatigue, which Rabinbach has identified and emphasized, was already evident in Mosso's studies on emotions, which conceived of emotions as mechanistic components of the body-as-machine.

When we enter the domain of the life sciences, we confront a concept of emotion that differs in significant ways from what we might conceive as emotion. Emotion was not a psychological category that pertained to actions and their motivations, to the subjective experiences of subjects or laboratory organisms, or to the overt expressions or behaviors of animals and humans, such as crying, laughing, or barking. Rather, emotion in physiology denoted a visceral event – a blush, a blood vessel 'spasm', a peristaltic wave, or the arrest of salivation. Thus, when we read in physiological texts that an animal or human had an 'emotion' during the protocol or the examination, we should not expect to hear a description of what we usually perceive of as an emotion, namely, various behavioral, gestural, or linguistic communications. Rather, we should seek to identify a visceral incident that occurred during the experiment or the clinical examination.

According to the Harvard psychologist John Gilbert Beebe-Center, the publication of William James's theory of emotions in 1884 gave a strong impetus to the development of physiological approaches in the *psychology* of emotions. A second important incentive was Wilhelm Wundt's 1899 appeal to physiological concomitants in support of his 'tri-dimensional' theory. This latter development was especially important for the investigation of hedonic tone (pleasantness and unpleasantness). By 1906 these investigations in psychology had reached their acme and began to decline, although they never disappeared.[5] The study of emotions in *physiology* began many decades prior to James's 1884 theory of emotions. It was already evident in the very early stages of modern physiology, in the late eighteenth-century studies of Albertus Haller. In his 1786 *First Lines of Physiology,* Haller analyzed the physiology of laughter, weeping, 'joyful' and 'sorrowful' passions, 'diarrhoea caused by fear,' pleasure and pain, 'palpitation in the heart,' 'venereal pleasure' in females, as well as in males, and more.[6] During the first two tiers of the nineteenth century, physiologists, naturalists, and clinicians – Xavier Bichat, Charles Bell, Johannes Müller, Thomas Burgess, Moritz Schiff, Duchenne de Boulogne, Claude Bernard, Elie de Cyon, and Charles Darwin – studied the emotions and their expressions from a physiological perspective. It was, however, only during the 1860s and 1870s that experimenters began to produce and measure emotions inside the laboratory – as objects of scientific knowledge.

The historiography of the modern study of emotions usually includes Charles Bell, Thomas Burgess, Pierre Gratiolet, Duchenne de Boulogne, and culminates with Charles Darwin. I focus on an alternative history, which began with Claude Bernard, Moritz Schiff, Elie de Cyon, Angelo Mosso, and W. P. Lombard. These two divergent traditions are not mutually exclusive, and converged in Darwin's *The Expression of the Emotions in Man and Animals* (1872). Darwin was well versed in the experimental studies of Claude Bernard on the emotions – and was, in his turn, widely read by physiologists, such as Angelo Mosso, Sir Charles Scott Sherrington, and Walter B. Cannon, as well as by psychophysiologists, such as the French Charles Féré. Crucial to these two distinct lineages – the former focusing primarily on facial expression, the latter on internal physiology – was the embodiment of emotions or their physicality, and material manifestations.

The embodied conceptualization of emotions has a long and convoluted history. Early twentieth-century physiologists invoked René Descartes' *Les Passions de l'âme* (1649) as the primary and direct source for their own physicalist pursuits.[7] The renowned late nineteenth- and early twentieth-century British physiologist, Sir Charles Scott Sherrington, pointed to the James–Lange–Sergi theory of emotions, which presented 'the extreme antithesis to the spiritualistic conception of emotion.'[8] The psychologist Carney Landis, who published widely on facial and physiological manifestations of emotions during the interwar years, observed that Charles Bell's 1806 treatise, *Essays on the Anatomy of Expression in Painting* (1806), was the direct forerunner of Darwin's and James's physiological approach, since Bell had insisted that bodily expression was a basic fact of emotion.[9] Contemporary authors like Ian Hacking have proposed yet a different genealogy, arguing that 'in the early modern period ... [through Hume and Kant] emotions were inner (like ideas), and the possibility of expressing them was a sort of happy coincidence, not part of their essence. Only as the ideas of Romanticism took hold did we come to think that the expression of emotions is part of their very nature.'[10] Others propose a longer genealogy, which dates to Aristotle's writings.

The embodied conceptualization of emotion was evident in many texts that studied the emotions during the second half of the nineteenth century – in Alexander Bain's *The Emotions and the Will* (1859); Guillaume-Benjamin Duchenne de Boulogne's *Mecanisme de la physionomie humaine ou de l'expression des passions applicable à la pratique des arts plastiques* (1862); Charles Darwin's *The Expression of the Emotions in Man and Animals* (1872); Carl George Lange's The Emotions: A Psychophysiological Study (1885); Alfred Lehmann's *Die Hauptgesetze des menschlichen Gefühlslebens* (1892); and in other major texts of the late nineteenth century. As Bain explained in summarizing the major thrust of the argument that would appear throughout the late nineteenth century, 'the concurrence of the moving organs is necessary to the full development of the feeling....What we take merely as signs of the emotion are a part of its own essential workings, in whose absence it would be something entirely different'.[11] Or as Charles Darwin explained, 'most of our emotions are so closely connected with their expression, that they hardly exist if the body remains passive'.[12]

This manner of conceiving emotions immediately implicated techniques for their physical – that is experiential – control and manipulation. It positioned emotions in the realm of the physical and lawful, rather than that of the spiritual and capricious.[13] What seemed to be capricious on the psychological side was law-like and causal from the biological–organic perspective. Ivan Petrovich Pavlov would make the exact observation, as did a whole generation of investigators. The materialization and mechanization of emotions went hand-in-hand with their animalization, and evoked resistances and reversions. Darwin, for example, had animalized practically all the expressions of emotions, yet he persisted in endowing the blush with a unique status. The blush was, according to Darwin, an exclusively human reaction. The experimenters who followed in Darwin's footsteps rejected what they perceived as Darwin's misguided aversion to animalize the blush. In his 1884 book, *Fear*, Angelo Mosso, who in many respects inaugurated the laboratory study of emotions, provided a physiological dissection of the blush. He deliberately discussed, recognized, and analyzed the 'blush' in his laboratory animals, and reduced the blush to simple mechanistic processes in the body-machine.

What is an emotion?

> Les changements dans la pression du sang résultant des expériences sont-ils causés par une activité purement intellectuelle, ou sont-ce des sensations produites par les excitations des sens, ou bien faut-il les considérer simplement comme les effets des émotions et du sentiment qui les accompangnent?[14] Friedrich Kiesow, 1895

> Heart rate, pulse rate, blood volume, and blood pressure have all been measured in experimentally arranged emotional situations by the use of devices called cardiotachometer, cardiograph, sphygmanometer, plethysmograph and sphygmomanometer ... .Too many organic events influenced in too many ways the cardio-vascular responses for them to serve as indicators of emotion.[15] Adolescent Study Unit, 1935

Before lunging into an examination of the specific practices for producing emotions, it is imperative that we understand how physiologists construed the notion of emotion. Modern students of emotions constructed a unique operational definition of emotion in their laboratories, a shop-floor definition that did not necessarily reflect various conceptual, theoretical, or hypothetical formulations of emotions. Emotion was now defined in terms of experimental manipulation – the operations necessary for its creation (such as isolation, purification, replication), – and graphic representation. Thus at the level of production, identification, and representation, William James's famous 1884 question – 'What is an Emotion?' – received a new and surprisingly uniform response from various investigators who studied emotions in different experimental/clinical settings, and who were often embroiled in protracted theoretical and disciplinary debates regarding the nature of emotion (for example, purposive versus destructive, peripheral versus central, social versus biological).

The operational or working definition of emotion was not a stable category, yet during the early twentieth century a number of its basic features crystallized and seems to have remained stable for several decades. In laboratory terms, emotion was the residual physiological processes that were observed in an organism after the exclusion of all other possible causes for the observed visceral changes. Emotion was defined in negative terms: it was not pain, nor physical or mental work, because these factors were differentiated from emotion in the laboratory. By eliminating 'non-emotion,' the experimenter isolated 'emotion.' Thus we find that laboratories studying the emotions – in the brain or elsewhere – developed practices for targeting 'non-emotions' and eliminating *them* from the protocol.

One such 'non-emotion' was pain. Though I do not intend to discuss in this chapter the complex relationships between pain and emotion, suffice it to say that emotions, feelings, and pain had existed in a dynamic, often confusing and overlapping relationship. During the late nineteenth century physiologists of emotion began to disentangle pain from emotion in their laboratories – not by formulating new arguments or theories concerning their distinct natures, but by devising manipulations that targeted and eliminated 'pain' from experiments on 'emotion'. The study of emotions demanded, in a sense, the elimination of pain. The particular conceptualization of an emotion – as not pain – was already present in various protocols from the late nineteenth century.[16] It was, however, best articulated in Walter B. Cannon's forgotten discussion of 'fesselungsdiabetes'. In his short discourse on the phenomenon, Cannon distinguished between 'fesselungsdiabetes' – a term coined by R. Boehm and F. A. Hoffman in the late 1870s – and *emotional* glyscosuria.[17] The details of his argument are not important here. What is important is that his distinction entailed the argument that physiologists had to distinguish between pain and emotion in interpreting physiological phenomena.[18]

The emphasis on eliminating pain in order to study emotion appeared in many protocols, though the particular maneuvers necessary for these eliminations depended on the specific details of each protocol. In order to eliminate pain, experimenters usually mentioned what might have seemed like trivial details to the uninitiated: the use of *small* cages; the employment of various types of barriers that prevented direct physical contact between stimulus (dog) and reactor (cat or rabbit); and the deployment of a type of practical benevolence, such as a 'comfortable holder' or the use of local anesthetics.[19] These practices underwrote the meaning of emotion, and were collapsed with the practices that directly sought to evoke emotions in animals (and humans).

Producing emotions

> A cat in a small cage was placed in front of a large dog. The dog tried to get at the cat, and being prevented by the cage, became more and more enraged. The cat inside the cage became more and more frightened. A sample of the blood of both animals was procured both before and after.[20] *New York Times*, 1931

> In a dark, quiet room, the investigators put the patient on his back. They passed a slender tube through one nostril into his stomach, so that a sample of the stomach fluid could be tapped at any time. They talked soothingly to him, urged him to relax and think peaceful thoughts. When he was in a good frame of mind they took a stomach sample. Then they began to talk with him about other things (with the tube through his nostril he could talk well enough) – unpleasant things, things that made him resentful, anxious, angry, frustrated. They continued their calculated tactlessness till his voice and manner showed that he was in a good dither. Then they took another stomach sample. The experience was painful but it served the cause of science.[21] *Time Magazine*, 1940

Imagine yourself as a naïve spectator on a fortuitous visit to one of various laboratories of emotions during the first three or four decades of the twentieth century. Depending on the exact moment of your intrusion, you might see a tumbler containing water, an egg, a motion picture, an alcoholic drink, a large old rat, real (or rubber) snakes, letters, frogs, pictures of skin diseases, a picture of human entrails, nude art studies, pornographic pictures, and the *New Testament* opened to Luke 6:18–49. Suddenly the room would resonate with the sounds of gunfire, firecrackers, laughter, shouting, barking, hissing, or the melody of a popular or technical piece. The experimenter in the room might read aloud passages from Barbusse's *Under Fire*, a list of words, nonsense syllables, or even the word 'socialist.' Extensive as this list of *emotional stimuli* might seem, it is greatly abridged. The complete inventory would, theoretically, contain the vast multitude of events in the natural and preternatural worlds, for the experimental protocol also specified that emotions could be conjured up 'centrally,' from within, personal memories or even private fantasies serving as laboratory *stimuli* – 'asking the observer to recall some occasion when an emotion was felt, to relive the experience and revive the emotion as completely as possible'.[22]

The major sites for the production of scientific emotions were the laboratory and clinic. The intra-laboratory and clinical pursuits of emotions depended on a constant and reliable source of emotions – an emotion generating technology – that would produce a steady stream of emotions. Emotions could not be stored, frozen, or preserved in their original form, and had to be produced on-site and on-demand. In human subjects the number of emotions that experimenters at least strove to investigate was extremely expansive, and included such emotions as joy, sadness, fear, disgust, anger, shame, embarrassment, frustration, and more. In animals, by contrast, only a very limited number of emotions were investigated. Research involving animals and emotions during this early period focused almost exclusively on rage and fear, with some experimenters including anxiety, joy, and jealousy.[23] Yet, it was the *physiological* study of *animal* emotions that would, ultimately, come to have a much more dominant and central position in the history of the study of emotions. As Julian Huxley explained, 'In birds the advance on the intellectual side has been less, on the emotional side greater: so that we can study in them a part of the single stream of life where emotion, untrammeled by much reason, has the upper hand.'[24]

The first step in every experiment on emotion was the production of a zero, or neutral emotional state. This preliminary condition of the mind served two basic purposes: it functioned as a baseline – shared, standardized and replicated across subjects – which ensured that each and every mind would begin the protocol from the same mind-state, and it represented the ideal starting point of any physiological investigation – a zero state from which all background noise had been eliminated.[25] In animals the creation of a zero state was achieved by new laboratory manipulations that were developed specifically in the context of research on emotions. Animals were 'petted' on the back, 'stroked' on their bellies, or held 'gently' on the keeper's lap in order to eliminate emotional excitement.[26] In rare cases, anesthetics were used on animals, as well as on humans, in order to purge emotions from the mind.[27]

The primary method of producing emotions in animals was, as one physiologist explained, the 'natural enmity between two laboratory animals,' the cat and the dog.[28] Or as another physiologist put it: 'the very common emotional response offered by a cat when confronted by a dog.'[29] The basic concept that governed the production of fear and rage inside the laboratory was what physiologists defined as 'sham fight[s].' These were struggles in which both cat and dog were made to enact the fight situation, generating in the process natural emotions of fear or rage, but in which neither would actually engage in 'real' physical fighting. The basic setup was to allow a dog to go through the motions of an attack while ensuring that the cat was never harmed or in actual contact with the dog. The protocol usually consisted of a cat 'placed, separately, in a small wire cage, and ... barked at by an energetic little dog, which jumped ... and made signs of attack'.[30] As George W. Crile wrote in his private notes, in order to produce emotional stimulation alone, 'rabbits were frightened by dogs, but not injured and not chased'.[31]

There were many variations on this basic scheme. On rare occasions, the experimenters themselves would initiate a 'sham attack' on the animal.[32] These 'sham fights' or 'attacks' often lasted for up to an hour, enabling the experimenter to perpetuate fleeting emotional states or measure the effects of long-term emotional exposures. By eliminating pain (cats were always protected), muscular exertion (cats were confined in small cages), and physical impact (cats were unable to run or flee), the physiologist isolated what he conceived of as a 'pure' emotion. Descriptions, such as 'rabbits were frightened by dogs, but not injured and not chased,' or rabbits were 'subjected to threatened attacks...by *muzzled* dogs' identified these experimental protocols as 'sham'.[33]

Less common techniques would involve tricking the animals by 'handling the starting switch [of the treadmill]...[this] produced emotional excitement with cardiac acceleration in the trained dog.'[34] Some physiologists, like T. R. Elliott, generated emotions by literally inducing a hallucination, using 'morphia' and creating what Elliott defined as 'hallucinations of dread'.[35] The selection of animals for these emotion-focused protocols was based on individual response and on a preliminary stage during which the experimenter interacted with each animal, in order to test its emotional qualities and quantities. 'Energetic reaction', 'readiness of response' and the frequency and ease of 'blushing' were important physiological criteria that physiologists never claimed to have standardized or replicated. As Angelo Mosso explained, 'I chose those animals [rabbits] that blushed most easily and frequently.'[36] J. J. Izquierdo, working in Walter B. Cannon's laboratory in the late 1920s, chose 'vigorous young cats, previously selected because of their energetic reaction to a barking dog'.[37] And Sydney William Britton selected cats 'for their readiness of response'.[38]

This basic *modus operandi* (laboratory-based interactive selection of individual animals according to emotional categories), also underpinned the selection of the *stimuli* – the dog. An 'excitable' or 'aggressive dog' was often selected as the preferred animal-as-stimulus for evoking emotions in the animal-as-object of study (a cat). Or as Burge explained, 'fortunately also there was in our possession a dog with a particular strong dislike for cats'.[39]

The nature of this selection process and the use of different dogs as stimuli might have been expected to arouse public or private concerns regarding replication and standardization. Yet these did not materialize at the level of the community, in spite of rare explicit remarks concerning the non-standard nature of these and other emotional stimuli. As Britton, working in Cannon's laboratory during the 1920s, observed in describing two different dogs, 'both were apparently of equal efficiency as exciting factors'.[40] L. Brouha had also reflected on the stimuli used in his laboratory, which were by 'no means quantitative'.[41] Or as Newton had put it in his study, 'Obviously no quantitative statement can be made regarding this phenomenon [fear] because the struggles [between cat and dog] themselves are not of uniform intensity.'[42]

Representations[43]

> It is interesting to be able to measure the power of emotion in terms so tangible as ounces of sugar.[44] Rollin T. Woodyatt, 1927

The *operational* consensus regarding emotions was also visible on the representational level. From the late nineteenth century new technologies for measuring and representing emotions emerged in laboratories that studied emotions. The early beginnings of instrument-generated graphic representations of emotions can be traced to the pioneering laboratory observations of the French physiologist, Claude Bernard, during the 1860s. Bernard applied Marey's new *cardiograph* to trace the heart during emotion in 1865.[45] By translating the heart's 'intimate functions' into omnipresent and permanent graphs on paper, Bernard literally 'read in the human heart'.[46] The slightest emotion, he explained, produced a reflex impression in the heart, 'imperceptible to all, except for the physiologist' and his instrument.[47]

Following in Bernard's footsteps was the Russian physiologist Elie de Cyon né Tsion. In his 1873 lecture, 'The Heart and the Brain,' Cyon described the relationships between the brain, emotions, and the heart.[48] Each emotion activated the heart; each emotion left its own 'peculiar and characteristic' curve on the graphic paper issued forth by the physiologist's new machine.[49] By the turn of the nineteenth century, measurements and representations of visceral reactions during emotions were studied in the laboratories of Angelo Moss, Charles Féré, Eugene Gley, Joshua S. Lombard, Alfred Binet, Charles Scott Sherrington, Friedrich Kiesow, Wilhelm Wundt, and others. Ugolino Mosso, for example, determined that his own rectal temperature increased by 0.7°C during a state of aroused emotions.[50] Francis Gano Benedict, future director of the Carnegie Nutritional Laboratories, measured the constituents of his own urine during the 'disturbances' wrought by exciting emotions in 1900–1901. And Charles Féré and Angelo Mosso recorded their colleagues' emotions inside the laboratory.

What did these experimenters quantify in an emotion? From the late nineteenth century to circa 1930 the most widely used method was to identify a peripheral physiological referent of the emotion, and to measure this referent as an indicator of the emotion. These peripheral concomitants of the emotion, which were 'habitually taken as outward expression of an inward feeling', were the only available physiological parameters that provided quantitative and objective information regarding the emotion.[51] Using these peripheral signifiers (heart rate, blood pressure, psychogalvanic reflex, saliva, and so forth), experimenters were able to determine the duration, pattern of appearance, pattern of dissipation, intensity, 'incubation period,' and even the quality of the emotion (some experimenters argued that specific peripheral physiological patterns indicated the type of emotion produced, but this was highly contested).[52]

The two basic techniques for producing visual representations of emotions, as I have argued elsewhere, were monitoring the body during emotion or sampling from the body's tissues and secretions during excitement. In the first of these techniques the organism was connected to one or more of several machines and continuously monitored on line. The output was a cinematic representation in which the flow of emotions, in their dynamic formation, undulation, and dissipation, was observed as it occurred in the subject's mind. Transcription was immediate: from body to instrument to representation. The variables that were measured in producing representations were relatively straight forward in physiological terms: blood pressure, pulse rate and shape, electrical resistance of the skin, respiration, blood flow to the limbs, and so forth. The output was usually in graphic form and the trained observer could translate in real time from transcription back to emotion. These on-line sessions lasted from a few seconds to a couple of hours and the transcriptions were produced in real time – as the experience unfolded.

In the second general technique the researcher sampled from the organism/human during its emotional experience and manipulated the sample – what one physiologist called the 'excited blood' – in the test-tube, long after the original experience had dissipated.[53] These samples – sweat, saliva, urine or blood – somehow preserved the experience and were converted into a representation of the original emotion (that had been present during the sampling). The most common procedure for manipulating these fluids was to characterize them biologically, chemically, or physically. The pH of a sample of saliva, the amount of glucose in voided urine, and the number of red blood cells from a phlebotomy were translated into levels of emotional excitement. Here, too, the final representation was usually in the form of a graph or numeric table and the emotion was often depicted in the form of a portrait, frozen and preserved in the space/time of a curve.

These clinical and laboratory representations of the body resonated with wider public conceptions. As one newspaper reported, the new 'Kiss-O-Meter' could measure objectively, through the body, the infatuation of two lovers during a kiss. The instrument, constructed by the Armour Institute of Technology, had been tested on various couples. The results showed a steady decline in the intensity of the kiss with age and years of marriage. The article also reported that 'ten husbands were persuaded to kiss their mothers-in-law. Here the highest that the machine was able to register was six. . .[on a scale of 0 to 120] and observers remarked that it might be interesting to add a minus scale to the dial.'[54]

During the interwar years, affect-gauging technologies spread into the judicial, commercial, clinical, entrepreneurial, governmental, private (romantic), and leisure spheres. The 'Lie Detector', 'Affectometer', 'Emotograph', 'Emotion-Meter', 'Stressometer', 'psycho-detectometer', 'Ego-Meter', and 'Kiss-O-Meter' were all conceptually similar devices that measured and represented emotions in numeral/graphical inscriptions. These various techniques for moving between experience and image shared in identifying the immediate and material body of the organism or human as the locus for acquiring knowledge and producing representations of emotion. Emotion as pure affective experience was not the material with which these different experimenters worked. Emotion was represented in the graphs as a pattern written in the language of the material elements of the body. A specific pattern of sweat secretions, a particular blood pressure curve, or a certain shift in the pH of the urine was read as 'emotion'. Thus the graphic and numeric representations of emotion never articulated emotion in units of experience, but in body units. In spite of the suppression of experience and the emphasis on the body, physiologists always maintained that their biological – body centered – investigations were studies of the experience of emotions. This argument was still relevant, even when experience itself was eliminated from some of the protocols that studied the physiology of emotions in animals.

Conclusions

The early investigators – Bernard, Schiff, Cyon, Darwin, and Mosso – undertook a most arduous task: to '[vivisect] the human heart' and to integrate and incorporate emotion into a mechanistically conceived body.[55] They created a machine that was imbued with emotion and mechanized the Victorian blush.[56] The laboratory of emotion was in a somewhat precarious conceptual and practical position. It was the ideal enclave for disciplining the passions (in the experimenters as well as in the objects of knowledge), yet it demanded and produced states of disorder – even loss of control – defined as 'emotion' by the experimenters.

The transformation of emotions into objects of *laboratory* knowledge entailed a series of moves and turns, which were essential for the incorporation of emotions into the culture and praxis of laboratory life. We can understand many of the subtleties of investigations on emotions if we keep in mind that emotion was redefined in new ways at the level of protocol, rather than explicit argumentation. These protocols, in their turn, were often implicated in broader political and disciplinary contexts. When physiologists defined emotion as 'not pain,' they avoided unnecessary political entanglements. The cats and dogs in their laboratories were not really attacking, or more importantly, really hurting, or violently injuring each other, but only *simulating* these acts of aggression. They were only 'sham' fighting. This elimination of pain from the laboratory was not only important for the study of 'pure' emotions – from a physiological perspective – but was also particularly significant in the late nineteenth and early twentieth century context of the antivivisection and animal protection movements. The elimination of pain was both a political and scientific act. In spite of this elimination of pain in the study of emotions, some critics viewed the study of emotions itself as entailing abuse, even if it avoided direct physical injury. As one correspondent wrote to the *New York Herald* in 1913: 'It is well known among interested persons that a monkey recently died of fright on a vivisector's table the result having been provoked but not desired by his tormentors…the arch vivisector and pride of the medical profession, Dr. George W. Crile. His extensive operations… [on] various species of animals are frequently discussed, but that he has deliberately frightened them to death is to some a new angle.'[57]

The exclusion and elimination of 'not emotion' – to isolate 'emotion' – was also embroiled in disciplinary boundary making between medicine and psychiatry during the early twentieth century. In the clinical realm, patients with *somatic* expressions of 'emotional' problems were transgressors – provocateurs – of the strict disciplinary boundaries between the body and the emotions.[58] These tensions involved the partitioning of the responsibility for particular symptoms between a psyche and a soma, between psychiatry and medicine. Psychosomatic clinicians struggled with the uneasy task of distinguishing between the two and proving emotional causation. The most common route for proving 'emotion' in the clinical context was, as in physiology, by eliminating other possible causes for the physiological pathologies that had been observed. Physiological interests were thus embroiled in disciplinary border making, in addition to science itself.

On a shop floor level the practices of the protocol re-inscribed broader social relationships. The relationships between experimenter and animal reflected the biological relationships between the cortex – as the center of rationality, knowledge, and control – and the sub-cortical regions of the thalamus or hypothalamus – as the centers of emotional expression and release. The experimenter, in the role of knower, controller, and rationalizer embodied the cortical convolutions. The animal, in the role of emotion, embodied the hypothalamus or thalamus. This embodiment was very real, for according to the reigning physiological model, during strong emotions the cortex-released control over the hypothalamus and the excited animal was literally and momentarily 'hypothalamic.' The scientists inhabiting these laboratories embodied the cortical characteristics of rationality and control. The laboratory of emotions was not only the space where animal encountered scientist, but also the space where embodied rationality encountered embodied emotion.

In concluding I would like to argue that the concerted study of emotions in the life sciences during the nineteenth and twentieth centuries challenges contemporary divisions in the history of science. Much of the modern study of emotions from the nineteenth century adopted a physiological perspective – even when it supposedly involved the psychological laboratory. Emotion, however, is often subsumed under the histories of psychology or psychiatry – even in our databases. This chapter reclaims emotions for the history of laboratory physiology and for clinical, non-psychiatric, medicine.

Notes

1. Angelo Mosso, *Fear*, trans. E. Lough and F. Kiesow 5th ed. (London; New York; Bombay: Longmans, Green & Co., 1896 [1884]), pp. 94–5.
2. Anon., 'How the Body Betrays the Mind', Science and Invention, *The Literary Digest* (1914), 68, pp. 153–5, 153.
3. Anon., 'Students Measure Fear by a Pupilometer, Kick Subject's Shins to Experiment on Anger', *New York Times*, 24 November 1925, p. 4.
4. Anson Rabinbach, *The Human Motor: Energy, Fatigue, and the Origins of Modernity* (New York: Basic Books, 1990), pp. 84–119.
5. John Gilbert Beebe-Center, *The Psychology of Pleasantness and Unpleasantness* (New York: D. Van Nostrand Co., 1932), pp. 37–46.
6. Albertus Von Haller, *First Lines of physiology* ed. Lester S. King, 2 vols. (facs. repr. New York: Johnson Reprint Corp, 1966 [1786]), vol. 1, p. 80.

7. René Descartes, *Les passions de l'âme* (Amsterdam: [Henry Le Gras] chez Louys Elzevier, 1649).
8. Charles Sherrington, 'Experiments on the Value of Vascular and Visceral Factors for the Genesis of Emotion', *Proceedings of the Royal Society of London* (1900), 64, pp. 390–403, 391. See also Harvey W. Wiley, 'Pepsin, Pep, and Patience', *Good Housekeeping* (1915), 61, pp. 805–9.
9. Charles Bell, *Essays on the Anatomy of Expression in Painting* (London: Longmans, 1806).
10. Ian Hacking, 'By What Link are the Organs Excited?' *Times Literary Supplement*, 17 July 1998, pp. 11–2.
11. Alexander Bain, *The Emotions and the Will* ed. Daniel N. Robinson (London: John W. Parker, 1859), pp. 8, 10.
12. Charles Darwin, *The Expression of the Emotions in Man and Animals* (New York; London: D. Appleton and Co., 1872), p. 237.
13. The radical reconceptualization of all emotions in naturalistic terms and the shift from a spiritual to an embodied-material representation were both also visible outside of the laboratory, as Steven Kern has recently argued in his study, *The Culture of Love: Victorians to Moderns* (Cambridge, MA; London: Harvard University Press, 1992). Between 1880 and 1920 literary and artistic representations of emotions underwent a radical shift from a spiritualistic depiction to an embodied, 'authentic' one.
14. Friedrich Kiesow, 'Expérience avec le sphygmomanometre de Mosso sur les changements de la pression du sang, chez l'homme, produits par les excitations psychiques', *Archives Italiennes de Biologie* (1895), 23, pp. 198–211, 198 (original emphasis).
15. 'Adolescent Study Unit, 1935', folder 7, box 17, YRG-37-V, Institute for Human Relations, Manuscript and Archives, Yale University Library. See also Christian A. Ruckmick, 'Emotions in Terms of the Galvanometric Technique', *British Journal of Psychology* (1930), 21, pp. 149–59.
16. The opening citation from Kiesow is also very indicative of the implicit distinction between intellectual activities, sensory activities, and emotional ones. In his paper Kiesow exposed subjects to various stimuli, for example, colors, and attempted to discover whether the changes in the plethysmograph were due to intellectual, sensory, or emotional responses to the artifacts. His experimental setup, however, was not as rigorous as Cannon's later experiments on animals.
17. Rudolf Boehm and Friedrich A. Hoffman, 'Beiträge zur Kenntniss des Kohlehydratstoffwechsels', *Archiv für Experimentelle Pathologie und Pharmakologie* (1878), 8, pp. 271–308, 300.
18. Walter Bradford Cannon, Alfred T. Shohl, and W. S. Wright, 'Emotional Glycosuria', American Journal of Physiology (1911), 29, pp. 280–7, 281–2; and Walter Bradford Cannon, *Bodily Changes in Pain, Hunger, Fear, and Rage: An Account of Recent Researches into the Function of Emotional Excitement,* 2nd ed. (New York; London: D. Appleton and Company, 1929), pp. 68–70. There was also a spate of papers concerning 'fesselungsdiabetes' in the German literature during the early 1910s. See for instance A. Loewy and S. Rosenberg, 'Über die normale Hühe des Blutzuckergehalts bei Kaninchen und Hunden', Biochemische Zeitschrift (1913), 56, pp. 114–6; Eckhard, 'Zur Deutung der Entstehung der vom vierten Ventrikel aus erzeugbaren Hydrurien',

Zeitschrift Für Biologie (1903), 143, pp. 407–40. It is obvious that some of these German investigators were familiar with Cannon's investigations.

19. See, for example, J. M. D. Olmsted and M. W. Morgan, Jr., 'Refraction of the Rabbit's Eye in the Unexcited and Excited State', *American Journal of Physiology* (1939), 127, pp. 602–4, 602; and J. J. Izquierdo and Walter Bradford Cannon, 'Studies of the Conditions of Activity in Endocrine Glands: XXIII. Emotional Polycythemia in Relation to Sympathetic and Medulliadrenal Action of the Spleen', *American Journal of Physiology* (1928), 84, pp. 545–62, 546. Physiologists as a group seem to have paid more homage in their published works to the manipulations necessary for eliminating the physical to produce the emotional. One simple explanation for why physiologists were more inclined to emphasize this particular point is that they did not have recourse to ideational stimuli (thoughts, hypnosis, words), as did psychologists. Some of the psychologists who employed physical stimuli also introduced various controls and emphasized the total elimination of the physical.
20. Anon., 'Find Rage and Fear are Fat Consumers', *New York Times*, 11 April 1931, p. 1.
21. Anon., 'Mind & Body', *Time Magazine*, 16 September 1940, p. 50.
22. 'Conference on Experimental Study of Human Emotions', 15 October 1926, p. 11, Div. A&P, Rec. Grp. DNRC: A&P 'Common Experimental Study of Human Emotions: Conference on Experimental study of Human Emotions: Third', October 1926, NAS-NRC Archives, Washington, DC. (these are the confidential minutes of a conference sponsored by the NRC).
23. In this essay I have focused on attempts to produce normal emotional reactions. In parallel with the experimental production of normal emotions in laboratories there were laboratories that focused on the production of emotional pathologies. These included the work of Horsley Gantt, Howard Liddell and Jules Masserman, and focused on animals.
24. Julian Huxley, 'Emotions in Feathers', *The Living Age* (1923), 317, pp. 418–25, 425.
25. It is important to emphasize at this early point that there were numerous and different reasons for producing emotions inside laboratories and clinics: as tests or as diagnostic procedures; clinical investigations that focused on the effects of emotions on health and disease; as part of the collective effort to standardize routine physiological measurements; and in the context of numerous theoretical debates and laboratory investigations that focused specifically on the nature of emotions. In this essay I will focus mostly on the latter group, but will also mention some of the techniques that were employed by other groups.
26. Walter Bradford Cannon, H. F. Newton, E. M. Bright, V. Menkin, and R. M. Moore, 'Some Aspects of the Physiology of Animals Surviving Complete Exclusion of Sympathetic Nerve Impulses', *American Journal of Physiology* (1929), 89, pp. 84–107, 92; H. F. Newton, R. L. Zwemer and Walter Bradford Cannon, 'Studies on the Conditions of Activity in Endocrine Organs: XXV. The Mystery of Emotional Acceleration of the Denervated Heart after Exclusion of Known Humoral Accelerators', *American Journal of Physiology* (1931), 96, pp. 377–91, 379.
27. In these new laboratories anesthetics were, then, tools for eliminating emotions, and not analgesics. The use of anesthetics was very problematic for a variety of reasons and was therefore limited. On anesthetics and emotions,

see H. I. Schou, 'Some Investigations into the Physiology of Emotions', *Acta Psychiatrica et Neurologica* Supplement (1937), 14, pp. 3–111, 82; and Frederick Hillis Lumley, L. B. Nice, 'Blood Sugar of Adrenalectomized Rats', *American Journal of Physiology* (1930), 93, pp. 152–5.

28. Cannon, *Bodily Changes*, p. 136.
29. Sydney William Britton, 'Neural and Hormonal Factors in Bodily Activity: The Prepotency of Medulliadrenal Influence in Emotional Hyperglycemia', *American Journal of Physiology* (1928), 86, pp. 340–52, 342.
30. Cannon, *Bodily Changes*, p. 72.
31. George W. Crile, 'The Kinetic Theory of Surgical Shock and Anoci-Association', St. Louis Medical Society, 15 Feb. 1913, St. Louis, Mo., 13, folder 82, carton 27, George W. Crile papers, MS 2806, Western Reserve Historical Society, Cleveland, Ohio.
32. Sydney William Britton and R. F. Kline, 'Emotional Hyperglycemia and Hyperthermia in Tropical Mammals and Reptiles', *American Journal of Physiology* (1939), 127, pp. 730–4.
33. George W. Crile, 'Studies in Exhaustion: III. Emotion', *Archives of Surgery* (1922), 4, pp. 130–53, 130; and idem, 'The Kinetic Theory of Surgical Shock'.
34. L. Brouha, S. J. G. Nowak, and D. B. Dill, 'The Role of the Vagus in the Cardio-Accelerator Action of Muscular Exercise and Emotion in Sympathectomized dogs', *Journal of Physiology* (1939), 95, pp. 454–63, 454.
35. T. R. Elliott, 'The Control of the Suprarenal Glands by the Splanchnic Nerves', *Journal of Physiology* (1912), 44, pp. 374–409, 384.
36. Mosso, *Fear*, p. 13.
37. Izquierdo and Cannon, 'Studies', p. 546.
38. Sydney William Britton, A. Hinson, and W. H. Hall, 'Neural and Hormonal Influences on Bodily Activity: Differential Factors Controlling the Heart Rate during Emotional Excitement', *American Journal of Physiology* (1930), 93, pp. 473–9, 474. For the emotional selection of 'some lively cats', see also John Fulton to Philip Bard, 19 January 1931, folder 190, box 14, Series I, John Farquhar Fulton Papers, Manuscript Group 1236, Manuscript & Archives, Yale University, New Haven. There was, presumably, a particular tension in these selections, because experiments on emotions required animals that were both capable of being emotionless and emotional. Thus the sought- after trait would have been emotional manipulability, rather than emotionality. However, in discussing their animals physiologists of emotions emphasized the selection of emotionally reactive animals and did not discuss the preferred selection of animals that could achieve and maintain the zero emotional level.
39. W. E. Burge and E. L. Burge, 'The Effect of the Emotions on the Catalase Content of the Liver', *American Journal of Physiology* (1917), 44, pp. 75–9, 76. See also F. A. Hartman, H. A. McCordock, and M. M. Loder, 'Conditions Determining Adrenal Secretion', *American Journal of Physiology* (1923), 64, pp. 1–34, 5; and W. B. Cannon, *Bodily Changes*, p. 55.
40. Sydney William Britton, 'Neural and Hormonal Factors in Bodily Activity: the Prepotency of Medulliadrenal Influence in Emotional Hyperglycemia', *American Journal of Physiology* (1928), 86, pp. 340–52, 343.
41. Le Brouha, Nowak and Dill, 'The Role of the Vagus', p. 454.
42. Newton, Zwemer and Cannon, 'Conditions', p. 380.

43. This section draws heavily on two previously published papers. Otniel E. Dror, 'The Scientific Image of Emotion: Experience and Technologies of Inscription', *Configurations* (1999), 7, pp. 355–401; and idem, 'Counting the Affects: Discoursing in Numbers', *Social Research* (2001), 68, pp. 357–78.
44. Rollin T. Woodyatt, 'Psychic and Emotional Factors in General Diagnosis and Treatment', *Journal of the American Medical Association* 89 (September 1927), pp. 1013–4, 1013.
45. Claude Bernard, 'Sur la physiologie du coeur et ses rapports avec le cerveau', in ibid., *Leçons sur les propriétés des tissus vivants*, ed. M. Émile Alglave (Paris: Germer Baillière, 1866), pp. 421–71. The cardiograph's principle was based on mechanical transmission: a stethoscopic drum was placed on the chest of the examinee, above the heart – the pulsation of the heart was transmitted from this drum and registered in the form of a curve.
46. Ibid., p. 437. On my use of 'translation', see Bruno Latour, *The Pasteurization of France*, trans. Alan Sheridan and John Law (Cambridge, MA: Harvard University Press, 1988).
47. Bernard, 'Sur la physiologie', pp. 469–71, 469.
48. Elie de Cyon, 'Le coeur et le cerveau', *Revue scientifique de la France et de l'étranger* (1873), 21, pp. 481–9.
49. Ibid., p. 487. For the quote, see Fernand Papillon, 'Physiology of the Passions', *The Popular Science Monthly* (March 1874), pp. 552–64, 559.
50. Ugolino Mosso, 'Influence du système nerveux sur la température animale: Recherches', *Archives Italiennes de Biologie* (1886), 7, pp. 306–40, esp. pp. 337–40; J. S. Lombard, *Experimental Researches on the Regional Temperature of the Head under Conditions of Rest, Intellectual Activity, and Emotion* (London: H. K. Lewis, 1879).
51. Charles S. Sherrington, *The Integrative Action of the Nervous System* (New Haven: Yale University Press, 1906), pp. 251–2. By the 1930s experimenters began to use the EEG, in addition to peripheral measurements.
52. There was a substantial literature on peripheral signifiers of emotions. See Hugo Münsterberg, *On the Witness Stand: Essays on Psychology and Crime* (New York: McClure, 1908); William Moulton Marston, 'Systolic Blood Pressure and Reaction Time Symptoms of Deception and Constituent Mental States' (Unpublished Ph.D., Harvard University, 1921); and Frederick Peterson, 'The Galvanometer as a Measurer of Emotions', *British Medical Journal* (1907), 2, pp. 804–6, 805. For 'incubation period', see Henry S. Upson, 'Emotions as Symptoms', *American Medicine* (August 1903), 6, pp. 228–9, 228. On the specificity of emotional states and physiological patterns, see Fernand Papillon, 'Physiology of the Passions', *The Popular Science Monthly* (March 1874), pp. 552–64; William James, 'What is an Emotion?' *Mind* (1884), 9, pp. 188–205; and idem, 'The Emotions', *Principles of Psychology* (New York: Henry Holt, 1890), Chapter 25. Both are in Carl Georg Lange and William James, *The Emotions* (Baltimore: Williams & Wilkins, 1922).
53. For 'excited blood', see Cannon, *Bodily Changes*, p. 43.
54. Young lovers usually rated between 100 and 110. Newspaper clipping, n.d., folder 'clippings', carton 7, John Augustus Larson papers, BANC MSS 78/160z, The Bancroft Library, University of California, Berkeley.
55. Angelo Mosso, *Fear*, p. 15.

56. Of all the expressions of emotions, the blush, according to Darwin, was the sole expression beholden only to humans. Mosso, however, disagreed with Darwin: 'blushing ... was not given to man as a sign of nobility ... .It is simply caused by the structure of our vital machine.' See Mosso, *Fear*, p. 16.
57. Susan Drake Bishop, 'Kill Conscience First', letter to the Editor, *Herald New York*, 23 December 1913.
58. James S. McLester, MD, 'Psychic and Emotional Factors in their Relation to Disorders of the Digestive Tract', *JAMA* (1927), 89, pp. 1019–20, 1019. This is part of a symposium on emotions and disease, in which papers were given by Charles Hugh Neilson, Rollin T. Woodyatt, Nellis B. Foster, and J. Ramsay Hunt followed by a discussion. Ibid., pp. 1013–25.

7

From Clever Hans to Michael Balint: Emotion, Influence and the Unconscious in British Medical Practice

Rhodri Hayward

The story of Clever Hans, the mathematical wonder horse, is now relegated to undergraduate textbooks on psychology and ethology. Yet for a few years, at the beginning of the twentieth century, people all over the world thought he was a truly remarkable beast. Owned and trained by Russian aristocrat, Wilhelm von Osten, he had been coached for ten years in elementary mathematics using a collection of skittles, carrots and an abacus. After two years of intense tuition, Hans had mastered basic numeracy, tapping out the correct answers to arithmetical problems with his hooves. A few years later he had graduated to more complex algebraic feats and his reputation as mathematical prodigy soon brought him to the attention of the Berlin Board of Education. Inspired by the pedagogical possibilities raised by Hans's tuition, the board established a twelve men investigative commission led by the Berlin psychologist, Carl Stumpf.[1]

Stumpf's initial investigations failed to provide any adequate explanation for Clever Hans's mathematical insights bar his equine genius. However, a second investigation by Stumpf's student, Oskar Pfüngst, revealed that the animal seemed to be responding to minute visual cues unintentionally generated by his audience and in particular by his trainer, Mr Von Osten. When the horse was blinkered and the questioner placed out of sight, Hans lost his mathematical insights. As Pfungst demonstrated, what had seemed to be an example of animal genius or possible telepathy was little more than a trained display generated by the unspoken expectations of his owner.[2]

As a model of the medical encounter, Clever Hans seems an unlikely candidate. Yet his unwitting simulation of mathematical prowess raised a problem which was to bedevil the doctor–patient relationship throughout the twentieth century. It implied that the emotions, behaviour and symptoms exhibited by the patient undergoing clinical examination might simply be artificial creations: artefacts generated by the unspoken expectations of the doctor just as Clever Hans's arithmetical performances had been generated by his owner's quiet acts of anticipation.

Like Clever Hans's mathematical performance, the interpretation and reinterpretation of the horse's behaviour rests upon a number of unspoken assumptions. It is contingent upon a specific model of influence and a tacit understanding of the mechanisms through which such influences might operate. Such models and mechanisms themselves reflect a wider set of social interests and political agendas.[3] In tracing the passage of influence and the possible origins of actions and ideas we engage in a forensic operation: assigning responsibility and performatively identifying the powerful agents and passive actors in a community.[4] This paper engages in a similar operation: it provides a sketch map of the changing concepts of influence in British medical practice, showing how the language of emotions transformed the doctor–patient relationship in the twentieth century.

The emergence of the historical unconscious

Ideas of influence, emotion and the unconscious were transformed in early nineteenth-century Britain.[5] Their contemporaneous modification was not coincidental, since the meanings of the concepts were interrelated and they each drew part of their significance and definition from each other. The arrival of mesmerism in England at the end of the eighteenth century saw the concept of the unconscious move from being a simple description of unwitting actions such as sleep and become attached instead to much more complex behaviour such as the entranced performances of mesmerised subjects. These mesmeric experiments had initially been seen as dramatic representations of the power of influence. In the demonstrations of physicians such as John Elliotson at University College Hospital and John Wilson at the Middlesex Hospital, the entranced subject seemed to succumb to the magnetic control of mesmerism. Their bodies obeyed unspoken requests, displaying in their flesh, miraculous cures and implanted anaesthesias.[6]

In these early experiments, 'influence' remained close to its etymological roots in astrology: it was seen as a magnetised fluid which flowed between the nervous systems of the mesmerist and his or her entranced subject.[7] At the same time 'unconscious' was understood as an adverbial description rather than as a reference to the independent psychical phenomenon we understand it to be today. And in contrast to our contemporary understanding, nineteenth-century models of both influence and unconscious lacked any fundamental association with the emotions or the passions. The transformation of influence and the reification of unconscious into its modern conception as a determinative emotional apparatus is contingent upon two factors: changes in the local practices used to elicit and imagine unconscious performances, and, as we shall see, larger problems surrounding the images and activities of the medical profession.

By the middle of the nineteenth century, it was possible to trace a significant shift in the understanding of 'emotion', 'unconscious' and 'influence'. As Thomas Dixon has shown, British philosophers and mental physiologists had, by the 1850s, begun to conceive of the emotions (as opposed to the passions) as biological states independent of the rational will.[8] At the same time, 'unconscious' was starting to be seen as an actual entity rather than as a state of being.[9] This transformation arose in part from changes in the form of unconscious performance: it moved from being an unwitting activity and instead, with the rise of trance speeches and clairvoyant displays, came to be seen as a hidden aspect of the individual, which was made apparent in the mesmerised state.[10]

This reification of the unconscious was not simply a linguistic operation. Rather it was sustained by larger changes in the methods and techniques deployed to bring about the mesmerised state. Whereas the traditional tools of mesmerism, the wooden wands and magnetised baths, had originally been deployed to bring about a state of rapport between the mesmerist and the subject, the new methods aimed to control expectation and attention.[11] In the writings of hypnotists, beginning with James Braid, the control of attention, its fixation on certain objects or physical processes, is seen as the key to achieving the physiological effects that had originally been established in the mesmeric rapport.[12] In the writings of medical observers from Henry Holland to James Paget and Daniel Hack Tuke, it is the sustained action of attention and expectation, which is seen as generating somatic changes previously witnessed in the mesmeric anaesthesia and the clairvoyant cure.[13] Holland recounted how the sustained observation of limb could bring on a spontaneous haemorrhage while Tuke documented in great detail the marvellous effects, from dramatic cures to apparent hydrophobia, which could be brought about through the sense of anticipation.[14]

The writings of these authors bring out the clear difference between nineteenth-century models of the unconscious and those deployed by the psychoanalytic establishment today. In contrast to our current understanding, which sees the unconscious as the repository of our repressed emotions, the unconscious in Victorian writings is seen as a product of our attentive anticipations, or our embodied future hopes. Perhaps the most dramatic example of this difference is seen in the literature surrounding the new condition of 'railway spine'. Although the various paralyses of railway spine are now often considered to be a kind of pre-cursor to post–traumatic stress disorder and a condition produced through the repressed terror of the locomotive accident, Victorian commentators understood it as an effect of the expectation of damage and pain.[15] As one doctor explained:

> Railway collisions, their sudden occurrence, their dramatic setting, association of large numbers of the injured, the social prominence of many victims, the wide publication of newspaper reports, and the growing importance of the financial claim created a lurid mental picture in the mind of the injured and indirectly affected the general public in such a way as to provide a fertile soil for nervous disturbance.[16]

In this nineteenth-century model, psychosomatic disorders – such as the hysteric's paralysis or the neurotic's dyspepsia–were generated by a set of infectious anticipations rather than repressed emotions.[17] The patient's body became a mirror of their own hopes and assumptions and their physician's ideas. The control of contagious expectations had been raised as a problem in Victorian investigations into spiritualism and telepathy but by the end of the nineteenth century this problematisation had moved from psychical research into medicine.[18] The extent to which the physicians and ideas and personality influenced the presentation of disease became an ongoing controversy, reaching its most technical exploration in the French psychiatric debates over the status of hysteria.[19]

The psychotherapeutic examination, many psychiatrists believed, could be seen as a prime example of the process of unwitting suggestion. Pfüngst's demystification of the Clever Hans phenomenon was echoed in the arguments used by Nancy School psychiatrists (Hippolyte Bernheim and J. A. Liébault) in their critique of Jean-Martin Charcot's work at the Salpêtrière.[20] Liébault and Bernheim claimed that the stereotyped behaviour exhibited by Charcot's hysterics was little more than a trained performance produced through the suggestible patient's unwitting surrender to the expectations of his or her physicians.[21]

These criticisms were publicised during the First World War by Joseph Babinski and Arthur Hurst. Babinski was a former pupil of Charcot's who later rejected his teachings. Hurst, a British gastroenterologist who in 1907 had studied in Paris with Babinski and Joseph Dejerine.[22] In their investigations into apparent cases of shell-shock, Hurst and Babinski had both shown that the hysterical anaesthesia that Charcot had so assiduously mapped in his Salpêtrière demonstrations were in fact artefacts dictated by the investigating physician's own manner of examination.[23] Thus Babinski saw the doctor as one possible source of pathological influence and psychogenic infection. In the official British military medical manual on hysteria, Babinski argued that

> Doctors themselves do not sufficiently realise the good or bad effects which their statements, comments, examinations and gestures may have upon the minds of their patients, whom suffering, privations and moral or physical commotions have made very responsive to suggestion, doctors, we say, sometimes unconsciously exercise a hetero suggestion which is all the more powerful as their profession gives them authority.[24]

This idea of the pathogenic presence of the doctor was widely adopted by members of the Royal Army Medical Corps during the First World War. Thomas Arthur Ross, the Medical Director of the Cassell Hospital recorded how the shell-shocked servicemen he was treating suffered an epidemic of apparent epileptic attacks when visited by a voguish London physician.[25] Likewise the pioneer medical psychologist Millias Culpin noted how his own use of electrical equipment in the investigation of limb injuries was likely to produce an anaesthesia or paralysis, even if the recording apparatus was turned off.[26]

These arguments were soon incorporated into a growing British critique of the emergent discipline of psychoanalysis. In his Croonian Lectures of 1919, Frederick Golla, pathologist at the Central Pathological Laboratory at the Maudsley Hospital, argued that the psychological associations produced by the analytic subject were not the product of any endogenous neurosis, but instead reflected the unspoken agendas of their psychotherapeutic interrogators. The analyst's intense focus on the patient's performance meant that the subtle flicker of his eyes, or the changing inflection of his voice, could elicit evidence of a classic sexual neurosis just as easily as a maths example could be produced in Clever Hans.[27] For Golla, the evidence produced in the psycho-analytic encounter held the same implication as the early telepathic experiments conducted by the Society for Psychical Research. The initial excitement over the apparent acts of thought-transference was revealed in these investigations as nothing more than the mind-readers' 'interpretation of signs, consciously or unconsciously imparted by the touches, looks or gestures of those present'.[28] This misplaced excitement over the accidental products of slipshod investigations was, Golla believed, the distinguishing feature of psychoanalytic research.

Golla's criticisms initially appear to be part of standard critique of psychoanalytic practice, ranging from the early evaluations of Bernard Hart to the sophisticated analyses of Mikkel Borch-Jacobsen, which has long argued that the discipline rests upon a series of philosophical exclusions, of which the most important is the question of suggestion.[29] And it is true that Golla himself can be firmly placed within a British psychiatric tradition, which engaged in a persistent and sniping critique of the work of Freud and his followers. Through his work at the Central Pathological Laboratory, he became the protégé of Robert Armstrong Jones, who had complained that psychoanalysis was equivalent to a kind of psychological grooming in which distressed minors were turned into sexual obsessives.[30] At the Maudsley he worked with Edward Mapother who would himself claim that the accession of Freud to a fellowship of the Royal Society was equivalent to the anti-Christ being granted a bishopric.[31] This is not the place to rehearse English psychiatric reactions to the advent of psychoanalysis – that story has already been well told in the work of Michael Neve and Trevor Turner.[32] Instead I want to explore in greater detail Golla's model of patient and illness, before considering how that model contributed to a new understanding of the doctor–patient relationship.

If, according to Golla, the psychotherapeutic encounter was little more than a circus sideshow, then the manifestation of psychical illness was little more than a display of amateur dramatics. In assessing psychiatric casualties returning from the Western Front, Golla used a galvanic skin response machine to measure the physical signs of apparently traumatic emotions. To his surprise, he discovered that those who exhibited the most dramatic forms of hysterical trauma also demonstrated the least significant change in electrical resistance or physical tone.[33] Rather than suspect any weakness in his instrumentation, Golla hypothesised that hysterics lacked the nervous energy necessary to maintain an emotional life and instead learnt to dramatise their lives as if they were experiencing their own overwhelming emotions. As Golla wrote:

> The *mise en scene* of the galvanometer room seemed to be extremely conducive to the exhibition of hysterical crises … .One young soldier

> suffering from a hysterical contracture of the foot broke down during an examination; tears rolled down his cheeks, he addressed his dead brother in language savouring of a South London melodrama, he asked why he himself had not been killed in his brother's place so that the favourite son might have been left to comfort his poor old father, and all the time whilst he wailed and wept the spot of light from the galvanometer mirror remained steady.

Thus Golla concluded that the hysterical patient must be

> an egocentric individual without strong or durable emotions though anxious to impress the outer world with the gravity and intensity of his experiences. Always more ready than a well bred person should be to impress his fellows, he will not willingly apply himself to any task involving strenuous exertion, fatigue or danger. The total personality conveys an impression of flimsiness. He appears to be a very different person to the volcano of ill suppressed sexual passions that some writers on psychotherapy have portrayed.
>
> The symptoms of hysteria convey to the observer a certain general impression of what can be called theatricality – though possibly the stage ill deserves such a comparison. The necessity under which the cinematograph actor finds himself conveying his mental states to the audience by emphasising all the mechanisms of expression has much in common with the conduct of the hysteric.[34]

Golla saw the therapeutic encounter as a form of theatre. A theatre in which the patients were driven to act, not by their own passions, but by their idea of the psychiatrist's desires. Within his schema, the patient's words and actions are robbed of any significance. Indeed in the process of illness and cure, the patient is seen by Golla as little more than a puppet animated by the desires of a ventriloquising physician.

This transformation of the meaning and significance of the patient's actions was in part made possible by the deployment of electrophysiological technology. The tracings of the smoked drum kymograph and the movements of the photogalvanic lamp were held up as an alternative form of emotional display which escaped the pretences and counterfeits of social intercourse and the conscious mind.[35] The celebration of the emotions as true marks of the authentic life threw everyday interactions and behaviour under suspicion. As a writer in the

Daily Express commented, the galvanic skin response recorder was a 'sinister invention, and one which may send the whole social fabric – which is, as you know, built up of skilful lies – crashing into dust. The mask we all wear will be of no use when a dancing spot of light can give us utterly away.'[36]

The electrophysiological investigation of hysteria raised two intimately related problems. First, the idea of the patient's dissimulation, the implication that their symptoms are merely a form of theatre; second, the problem of the infective presence of the physician: the idea that the therapist's behaviour and the patient's expectations might together determine the form and presence of the disease.[37] To the extent that these problems were acknowledged by British general practitioners, they were generally approached using methods that had been pioneered by continental psychotherapists. There was a general attempt to trace the ownership or authorship of functional symptoms and apparent emotions away from the personality of the physician and onto the life history of the patient. Such a manoeuvre was quite easy to manage since it had already been developed in Victorian psychiatric works as well as popular novels.[38] As these works argued, hysterical illnesses clearly originated in the sufferer (rather than the investigating physician) since they could be shown to serve the patients' own selfish interests.

A good example of the ways that illness could be used to advance the plot and narrative agenda of warring characters was provided by May Sinclair's *The Helpmate,* published in 1908. Sinclair was a popular if slightly eccentric novelist, and her book explores the sexual tribulations of a young couple, Anne and Walter Magendie, whose honeymoon has been ruined by Walter's confession of a previous liaison with one Lady Cayley.[39] The news drives Anne Magendie into a permanent migraine, which leads to the suspension of their physical relationship. Walter discovers that the only way he can elicit any tenderness or physical contact from his wife is through his own bouts of influenza, although this concession is withdrawn once their child, conceived on the first night of the honeymoon, sickens and dies.

Feeling rejected, Walter enters into a series of extra-marital affairs, meeting again with Lady Cayley, although the strain of his infidelity provokes a new nervous collapse and a series of brain lesions. As his wife returns to nurse him out of his unconscious state she realises that her sexual abstinence has forced him into infidelity. Thus in this novel competing illnesses bring about the suspension and resumption of sexual relations. And the triumph of Walter Magendie in the sexual struggle with his wife lies in the pathological triumph of general paralysis over migraine.[40]

This reading of functional illness as a form of 'secondary gain' – to use the Freudian term – was to become the dominant approach among British general practitioners during the interwar period.[41] It was applied widely, being used to criticise the behaviour of patients and politicians. The gastroenterologist Sir Robert Hutchison's portrait of the 'abdominal woman' whose 'incessant demand for sympathy makes [her] a veritable vampire sucking the vitality out of all near her' was widely cited by other doctors.[42] At the same time, psychologically and politically minded practitioners, such as Frank Grey and Hugh Crichton-Miller, claimed that illnesses of Edward VII and Lloyd George were used as foils to avoid episodes of political embarrassment, with the king developing a convenient bout of appendicitis following Lord Salisbury's threatened resignation.[43] Such narratives, while discrediting the patient's behaviour, also possessed 'tertiary gain' for the doctor insofar as these stories undid any idea that the symptoms might be produced through the infectious presence of the investigating physician.

Although few general practitioners were doctrinaire Freudians, illness was largely understood as arising from an infectious episode in the patient's past. The most common approach drew on the work of Alfred Adler and Georg Groddeck, particularly their idea that illness could be seen as a way of retreating from personal failures and perceived inferiorities.[44] These ideas were promoted in England by Francis Graham Crookshank, a general practitioner from Barnes, South London who held paediatric posts at St Mark's, Belgrave Children Hospital and Hampstead General.[45] He had served as the Medical Director of the Hôpital Anglais Militaire in Caen during First World War and it was easy to imagine how his investigations into the malingering soldier may have later informed his treatment of the school-shy child. Yet his conversion to Adlerianism was largely brought about through his friendship with Dimitri Mitrinovic, the mystical Serbian émigré, pan-Europeanist and founder of the New Britain movement.[46]

Like Mitrinovic, Crookshank was not afraid of deeply speculative grand theories and today is probably best remembered for his eccentric work, *The Mongol in our Midst* (1924), in which he argued for a polygenist model of descent: a process of competitive evolution in which mongoloid and gorilloid races mix unhappily with their European counterparts.[47] He claimed that the presence of mongoloid blood could be seen everywhere from the crowds at King's Cross Station to Leonardo da Vinci's Virgin of the Rocks.[48] However despite the apparent eccentricity of his views, Crookshank was widely respected and his psychological theories were championed by Walter Langdon Brown, Lord Horder and the champion of basic English, Charles Kay Ogden.[49]

Following Adler, Crookshank argued that the illness should be seen as response to individual emotional frustration – a response, which was coloured by the particular physical weaknesses of the sufferer.[50] These physical weaknesses could either prompt new levels of activity as the patient strove to compensate for their individual failings or alternatively encourage and legitimise a defeatist attitude.[51] As Crookshank argued in his Psyche Miniature on *Migraine*: 'physical defect (or organ inferiority) offers the excuse for organisation of a "functional illness," and is as eagerly seized upon as the peg whereon is hung the vestment we wish to discard'.[52]

As Crookshank's friend, the Regius Professor of Medicine at Cambridge, Walter Langdon-Brown, noted, Adler's ideas were appealing to British doctors because they granted a constitutive role to the body in the original formation of a psychological illness – so that physical pathology and psychoneuroses together determined the course and progress of a disease.[53] Moreover, this somatic element in the constitution of psychological illness allowed general practitioners to develop, as Adler had, a set of diagnostic short cuts which rendered unnecessary the long and drawn out work of psychological analysis. Drawing upon Adler's theory of organ inferiority, Crookshank argued that patients manifested their psychological distress through a common somatic language or 'organ jargon': 'the migrainous man has astigmatism and practices coitus interruptus; the vertiginous man has some middle-ear catarrh and smokes too much, as well as being in fear of falling, physically and morally; the dyspareunic woman has a misplaced uterus and a husband whom she hates; the aerophagic doctor bolts his meals, is in debt, and has a flabby abdomen; while the subject of extra-systoles and palpitation drinks too much tea, has a congenitally poor heart, and masturbates'.[54]

Such diagnostic short-hands were not limited to the English Adlerians. The ophthalmologist William Samuel Inman, who worked closely with John Rickman and Millais Culpin and had undergone a training analysis with Sandor Ferenczi, drew upon Freud's early ideas of an organ language to argue that there was a persistent correlation between styes and worries about childbirth, squint and childhood rebellion, and the onset of cancer with personal disappointment.[55] These attempts to trace the origins of illness back onto the imagined emotions and life history of the patient could never successfully exclude the doctor as a pathogenic influence, since doctors themselves usually played such a significant role in the patient's story. Alongside the psychiatric critique of the doctor's behaviour as an unwitting source of infectious suggestions, more psychoanalytically inclined practitioners argue that patient's attachment to the doctor could also undermine the possiblity of cure. Henry Yellowlees, the Superintendent at the York Retreat who maintained a sympathetic interest in Freudian theory, complained of the 'psychosexual dependence' which developed in patients who surrendered themselves to the physicians' suggestive influence.[56] At the same time, Thomas Ross argued that these patients were likely to use their illnesses as part of a strategy to maintain the doctor's interest. Ross wrote

> A more serious thing is, which may prevent the disappearance of symptoms, is that the patient leans too much on the doctor. This leaning may only the natural leaning of the weak on the strong, or it may be due to pleasure in being in his society which may amount to being in love with him.[57]

By the 1930s, such ideas were common currency. The Welsh nature mystic, Margiad Evans, in her remarkable cystitis memoir, *The Wooden Doctor* (1933), relates in great detail how the course of her illness was bound up with the attraction she felt towards her general practitioner. She connects the onset of her infection to the initial affinity she felt towards him, and eventually recognises that his many attempts to treat the disease have been thwarted by her desire for him. The cystitis forces the doctor into activities which would normally be the province of a particularly brutal lover, but as Evans recognises he refuses to offer the one thing that could remove her suffering – his personal commitment to her. The novel ends with no hope of a cure in sight.[58]

The distaste which many doctors felt when confronted with the effects of their personal influence in the bodies of their patients is apparent in the shifting models of persuasion and cure which developed throughout the interwar period. At one level there was a general fear of anything approximating to a direct personal relationship between doctor and patient. Although modernists like Henry Brackenbury, the vice chairman of the British Medical Association, might insist that the new holistic emphasis of medicine had revealed that 'the relationship between doctor and patient is not merely between two persons but two personalities'; more cynical commentators, such as Hugh Crichton-Miller, the director of the Tavistock Institute, noted that few doctors would attribute a cure directly to the effect of their personality on that of a patient, although many were happy to subject the patient to their own will under the disguise of placebos or pills.[59]

Many psychologically minded doctors seem to have agreed with William Halse Rivers that the goal of therapy was to undo any level of intimacy in the doctor–patient relationship and instead induce a new level of independence and self-reliance in the patient.[60] If, as M. O. Raven suggested in 1932, the doctor imposed the pattern of his personality upon the patient's illness just as the artist imposed his or her personality upon canvas, the finished work might be a sad indictment not just of his artistic or clinical skill but of his own pathogenic identity.[61] Certainly by this time there was a growing awareness of the pernicious effects the unwitting physician might have on the suggestible patient. Maurice Cassidy, in a lecture to students at St Thomas's in 1938, jokingly suggested that in the interests of public health the medical profession should be abolished, since despite its useful work on ruptures, fractures, malaria and syphilis, the net effect of medical research had been to increase the anxiety and suffering of the population as they repeatedly measured their blood pressure and palpated their torsos for signs of a dropped stomach.[62]

In response to these problems, doctors developed a number of practical and conceptual strategies. At a prosaic level, there was a new emphasis upon managing the information available to the suggestible patient. Roger Bannister, a Cambridge psychologist and Papworth consultant, urged doctors to develop the skills of poker players so that they would not broadcast unwitting hints.[63] Likewise medical educators such as Ralph Noble warned of the pernicious effects of overheard conversations on ward rounds and the *Lancet* reported on the suicides of curious patients who had surreptitiously examined their medical records.[64]

By the 1930s the medical dangers of anxiety were seen as legitimating the doctor's decision to withhold the truth from the patient. As Hugh Crichton-Miller and Walter Langdon Brown argued, the physical consequences of fear and pathological effects on the body of the overstimulation of ephedrine or adrenalin meant that the irrational demands of the patient's physiology would have to override the doctors ethical commitment to truth and honesty.[65] These practical measures were accompanied by a general reconceptualisation of the idea of influence. The old mesmeric model had depicted influence as something akin to a bacterial infection: it involved the transfer of magnetic fluids, or later, unwitting signs, which generated new forms of behaviours and presentations of illness in the patient. Yet alongside the bacteriological model of the process, there also existed older medical metaphors, which had seen influence operating as a kind of miasma; with the general environment, rather than individual encounters shaping the person's character and destiny. As Robert Douglas-Fairhurst has shown, this miasmatic conception of influence had enjoyed considerable currency during the Victorian era, with the Carlylean idea of the psychological atmosphere or ethos being seen as the determinative element in men's lives.[66]

Moving between different models of influence served a forensic function. It shifted the authorship of symptoms away from the influential physician onto the general environment. It was a rhetoric, which freed the psychologically minded physician from the idea that their actions were solely responsible for the generation of the illness. Certainly it is very striking that the same Adlerian physicians who advocated a general recognition of the psychological factors in medicine were at the same time engaged in a general interrogation of the bacteriological model of disease. Crookshank himself led this wave of anti-bacteriological scepticism.[67] Allying himself with a version of William of Ockham's nominalism, Crookshank argued that

> We have no right to speak as if *diseases* were the primary objects of our attention. Diseases only exist subjectively in the mind like *Virtues* or *Vices*: they are fictions we have invented for our special purposes. The 'clinical entities' are the patients themselves who are ill in different kinds of ways and for different reasons. Particular cancers and broken legs and fatty livers have real existence of course, just as have petty mouths and ugly noses. But Cancer, Influenza, Pneumonia and so forth, have no greater claim to be called *entities* or objects in Nature with fixed characters, than have Beauty, Truth, Murder and Lust.[68]

This was a position he was to reiterate in his massive edited work on the influenza pandemic of 1919 and in his appendix to Ogden and Richards's *The Meaning of Meaning* (1923): bacteriological medicine, he argued, rested upon confusion between words, ideas and things.[69]

As David Cantor has shown in his essay on 'Neo-Hippocratism in Interwar Medicine' Crookshank's position was shared by many elite commentators: Walter Langdon-Brown, John Ryle, Alexander Cawadias and Sir Robert Hutchison all questioned the realist language of the laboratory with its confusions of actual entities with hypothetical explanations.[70] It was also a position supported by many psychiatrists, like Arthur Brock and Edward Mapother, who criticised psychoanalysts for their conflation of descriptive categories such as 'unconscious action' with the real objects of biology or physiology.[71]

In place of the bacteriological model of infection or individual influence, these doctors tried to map illness onto a wider political and cultural environment. In his work on influenza, Crookshank championed the Hippocratic model of 'airs, waters and places' as causes of disease and it was a short step to provide a psychological or cultural gloss on this model. Anxiety and illness were seen as the products of a cultural miasma. Thus among Crookshank's collaborators, Robert Dick Gillespie at Guys and Hutchison at the London Hospital, both complained that the age had succumbed to 'hypochondria epidemica'. Hutchison grumbled that the modern physician was now besieged with health faddists, nudists, dress reformers, stool collectors and the anxious parents of antiseptic babies who had all developed a new disease consciousness engendered by the pharmaceutical market and the Workmen's Compensations Acts.[72] Similarly Gillespie drew a parallel between the present nervousness and the previous wave of hypochondria in the eighteenth century. He complained that

> even the daily press is morbidly preoccupied with health: not merely advertisements for patent medicines but the lubrications of 'Harley Street doctors' fill the columns of the popular papers. The eighteenth-century hypochondriac developed a passion for simplicity and sought to emulate the noble savage in his unsophisticated ways ... .Today we are again exhorted to imitate him; but this time with the movements of our bowels, which (so we are told) should occur as with the savage, three times a day, after meals.[73]

And later, in his book on *Mind in Daily Life* (1933), Gillespie argued that modern physiology itself was engendering mental illness by making patients aware of the somatic effects of their personal anxieties and hence adding to their reasons for worry.[74]

In this idea of a cultural miasma, a number of themes are joined together. First, the nominalist critique of medical language, championed by Crookshank and his followers, is lent a new urgency, since the realist's conflation of words and things leads to new pathological forms of health anxiety. As Thomas Ross wrote in his Morrison Lectures: 'The habit of inventing diseases and the cognate one of attributing to real but trivial physical abnormalities the power of producing widespread symptoms is probably one that keeps up ill health more than any other mistake which doctors make.'[75] Second, as Christopher Lawrence has pointed out in his work on medical holism in inter-war England, this idea of a pathogenic environment allowed patrician physicians to engage in a form of social critique.[76] Authors like Thomas Horder, Cyril Percy Donnison and Langdon-Brown argued that modern civilization brought with it new forms of illness and nerve strain.[77] Third, such models undid the idea that the individual physician might be the source of the psychosomatic complaint by redirecting attention to the wider effects of culture and society.

This process of insulating the physician from the patient was reinforced by the adoption of alternative understandings of the curative influence. Against the scientific idea of cure as a form of direct intervention, many of the Adlerian writers of the 1930s sought to return to the theological model of the physician who influences through the power of their model character.[78] Late Victorian writers, such as the Metropolitan Police Surgeon and sometime demonologist Arthur Taylor Schofield, had earlier claimed that the moral example of the physician had an immediate reformatory effect on the patient's life.[79] And this idea of the exemplary character seems to be have been sustained by elite physicians such as Sir Dyce Duckworth, who in his Harveian Oration claimed that the fellows of the Royal College of Physicians were knit together in the image of Harvey and held up the form of his character to influence and restore the sick lives of their patients.[80]

In the 1920s and 1930s, this emphasis on the moral status of the physician began to reappear in the language of the English Adlerians. Crookshank, in his small book on *Diagnosis and Spiritual Healing* (1927), argued that the individual's innate capacity for repair could only be released if the sick were to encounter the 'objective personality of a physician...a personality marked by an ethical or moral superiority'.[81] This personality was effective, not because of its magnetic power as the Victorians had thought, but because of its ability to inspire a therapeutic emotional response in the patient. Crookshank's position was shared by many of the Adlerians like Charles Maurice Bevan Brown and Langdon-Brown as well as some of the more eclectic Freudians like MacDonald Laddell, Ian Suttie and Eric Graham Howe.

Eric Graham Howe was probably the foremost propagandist for this position. An aristocratic theosophist and Honorary Physician at the Tavistock, he had set up one of the first post-graduate psychotherapy courses for general practitioners.[82] In his lectures he repeatedly raised the problem of the physician's personal sense of inferiority and the dangerous effects of the various defence mechanisms, which might be used to disguise these feelings of inadequacy during the process of treatment.[83] Examples of this (according to Howe) included the doctor acting as if he or she was omniscient or the prescription of panaceas which may do more harm than good. The doctor's emotional state inspired a corresponding response in the patient. The magnetic influence which passed between mesmerist and entranced subject had been replaced by a new traffic in emotional signs and physical reactions.

By the early 1930s, this idea of the doctor as a therapeutic standing example was being explored in popular fiction. In her novel, *Private Worlds*, published in 1934, Phyllis Bottome (an enthusiastic champion of Adler's work), described the effect of staff relationships on the patients of modern English mental hospital.[84] In contrast to previous examples of mad literature which had concentrated on the eccentricities of individual lunatics, Bottome focussed upon the minor complexes and frustrating anxieties of a small number of doctors and their families as they struggled to overcome habitual defences and achieve fulfilling relationships. As she makes clear, the patients can only begin to recover once the doctors have themselves achieved personal happiness. Bottome's work won a wide readership: it was the seventh most popular work published in 1934 and a year later was released as an Oscar winning film.[85]

Bottome's novel does, I think, point to a wider transformation brought about by the change in models of influence. It turns the work of suggestive therapeutics from a kind of psychical engagement with the expectations of the patient into a new moral project in which the physician perfects his own character to ensure its efficacy as a curative agent.[86] This moral project overlaps with the earlier work on the pathogenic potential of the physician's personality to create a situation in which the doctor–patient relationship and the patient's illness are both imbued with a new significance. The course of the patient's illness under treatment was no longer a reflection of his or her psychological history – it also provided a surreptitious assessment of the GPs emotional life.

In the changing models of emotion and influence deployed in British general practice, we can trace out the emergence of deeper and more complex understandings of power and subjectivity in the doctor–patient relationship.[87] In early British psychotherapeutic and anti-psychoanalytic writings, the idea of suggestion was used to police the physician's behaviour: it highlighted the danger of methods of interview and diagnosis which might generate artefactual forms of psychological and psychosomatic distress. By the time of the Second World War the emphasis had changed, and suggestive threats of illness or the promise of cure seemed to arise not so much from the physician's behaviour, but from the actual structure of his personality.[88] It was not his or her conscious demands but their unconscious emotions which guided the development of the sickness. By 1948, the doctors who gathered around Michael Balint as part of his Tavistock seminar on problems in general practice understood the patient's illnesses or abnormal patterns of attendance as symptoms of their own failings as physicians.[89] Balint argued that the doctor possessed what he called 'an apostolic function' in which he attempted to win the patient over to his own understanding and in the process organised the patient's symptoms into a fixed pattern that would determine the trajectory of their illness.[90] This 'apostolic zeal' or overpowering influence arose in part as a defence mechanism, allowing the GP to avoid the pains of self-examination.[91] Similarly, the therapeutic use of reassurance by the doctor was seen as another form of defence, in which the GP avoided the humiliation of having to acknowledge his ignorance or impotence.

Balint claimed that 'every doctor willy-nilly creates a unique atmosphere by his individual ways of "practising" medicine and then tries to convert his patients to accept it'. He argues that the effects of influence can no longer be avoided. As he said: 'Whatever he does he cannot fail to influence his patient, and these influences will add up in the long run'.[92] Within the terms of competing models of influence, what is interesting about Balint's work is that he fuses two approaches: the atmospheric infection and exemplary cure, which were originally deployed to insulate the physician from patient and illness, and through this conjunction transforms them. The miasmatic influence of the doctor's personality engulfs the patient, and the patient's body and behaviour become symptomatic of the physician's own inadequacies and anxieties.

Through the use of these changing models of influence we can see how the relationship between psychology and general practice has been transformed. It has moved from being a project engaged with the management of the suggestible patient to become something akin to an old moral discipline in which the doctor perfects his personality in order to maintain his status as a therapeutic instrument.[93] It is a change that appears to have been generated through the very specific problems of medical science and primary care. The development of new concepts, which problematise the behaviour of the patient – the unconscious, suggestion, anxiety and transference – ends up infecting the work of the doctor. What had been pictured as a kind of discrete bacteriological transfer of unwitting signs became by the 1950s a miasma of unspoken infectious emotions. The language of the emotions which emerged in the twentieth century problematises every aspect of the relationship between doctor and patient from their life histories to their current behaviour. The magical aspect of emotion was that it made everything relevant and everything open to intervention.

Like the magic practised by the mesmerists and thought–readers, the magic of emotion was dependent upon a sleight of hand. Emotion, like expectation before it, was a purely performative concept. It referred to nothing ostensible, rather it knitted together certain aspects of the world – verbal descriptions, physiological states and social behaviour – through the mere act of enunciation.[94] These were deeply effective incantations. As the wide literature on placebo has shown, naming emotions and naming expectations generated new patterns of illness and health, which would transform the human body and the wider world.[95]

Notes

1. On Stümpf (1848–1936), see Carl Murchison, *A History of Psychology in Autobiography,* 2 vols. (Worcester, MA: Clark University Press, 1930), vol. 1, pp. 389–441.
2. Oskar Pfungst, *Clever Hans (The Horse of Mr Von Osten)*. A History of Experimental Animal and Human Psychology (New York: Henry Holt, 1911).
3. Roger Cooter, 'Anticontagionism and History's Medical Record', in Peter Wright and Andrew Treacher (eds), *The Problem of Medical Knowledge: Examining the Social Construction of Medicine* (Edinburgh: Edinburgh University Press, 1982), pp. 87–108.
4. On the politics of origins narratives, see Bruno Latour and Shirley C. Strum, 'Human Social Origins: Oh Please, Tell Us Another Story', *Journal of Social and Biological Structures* (1986), 7, pp. 169–87.
5. On emotion, see Thomas Dixon, *From Passions to Emotions: The Creation of a Secular Psychological Category* (Cambridge: Cambridge University Press, 2003). On influence, see Robert Douglas-Fairhurst, *Victorian Afterlives: The Shaping of Influence in Nineteenth-Century Literature* (Oxford: Oxford University Press, 2002). On the unconscious, see Henri Ellenberger, *The Discovery of the Unconscious* (London: Fontana Press, 1994 [1970]); Jonathan Miller, 'Going Unconscious', in Robert B. Silvers (ed.), *Hidden Histories of Science* (London: Granta Books, 1997), pp. 1–35.
6. Adam Crabtree, *From Mesmer to Freud: Magnetic Sleep and the Roots of Psychological Healing* (New Haven: Yale University Press, 1993), Chapters 7 and 8; Roger Cooter, 'The History of Mesmerism in Britain: Poverty and Promise', in Heinz Schott (ed.), *Franz Anton Mesmer und die Geschichte des Mesmerismus* (Stuttgart: Franx Steiner, 1985); Alison Winter, *Mesmerized: Powers of Mind in Victorian Britain* (Chicago: University of Chicago Press, 1998), Chapter 3.
7. OED.
8. Dixon, *Passions*, Chapters 5 and 6.

9. They drew upon the work of German authors such as Carl Gustav Carus and Eduard von Hartmann. See Lancelot Law Whyte, *The Unconscious before Freud* (London: University Paperbacks, 1966).
10. Again, this idea of mesmeric practice as a means of access to some occult aspect of the individual was developed in German occultism and popularised in Britain by authors like Catherine Crowe and William Howitt. See Catherine Crowe, *The Night-side of Nature* reprinted with an introduction by Colin Wilson (Wellingborough: Aquarian, 1986 [1848]); Joseph Ennermoser, *The History of Magic,* trans. William Howitt (London: H.G. Bohn 1854). On early British experiments in clairvoyance, see Joseph W. Haddock, *Somnolism and Psycheism* (London: Hudson, 1849).
11. On the rise of attention as an object of scientific investigation, see Michael Hagner, 'Toward a History of Attention in Culture and Science', *Modern Language Notes* (2003), 118, pp. 670–87; Jonathan Crary, *Techniques of the Observer* (Cambridge, MA: MIT Press, 1990).
12. On Braid, see James Milne Bramwell, 'James Braid, Surgeon and Hypnotist', *Brain* (1895), 19, pp. 90–153; Alan Gauld, *A History of Hypnotism* (Cambridge: Cambridge University Press, 1995), pp. 279–88.
13. The physiological rationale for this was provided in the work of William Benjamin Carpenter and Thomas Laycock, cf. Carpenter, *Mental Physiology,* 7th ed., (London: Kegan Paul, Trench, Trübner and Co., 1896 [1874]), Chapters 14 and 19, especially pp. 684–88.
14. Henry Holland, 'Effects of Medical Attention on Bodily Organs', in *Medical Notes and Reflections* (London: Orme, Brown, Green and Longmans, 1839); Daniel Hack Tuke, *The Influence of the Mind on the Body* (London: J. & A. Churchill, 1884).
15. Ralph Harrington, 'Railway Spine and Victorian Responses to PTSD', *Journal of Psychosomatic Research* (1996), 40–1, pp. 11–4.
16. William Thorburn, 'The Traumatic Neuroses', *Proceedings of the Royal Society of Medicine* (1914), 7, pp. 1–14, quoted in Ben Shephard, *A War of Nerves* (London: Cape, 2000), p. 16. For overviews of railway spine, see Thomas Keller, 'Railway Spine Revisited: Traumatic Neurosis or Neurotrauma', *Journal for the History of Medicine and Allied Sciences* (1995), 50, pp. 507–24; Ralph Harrington, 'On The Tracks of Trauma: Railway Spine Reconsidered', *Social History of Medicine* (2003), 16, pp. 209–23.
17. This model is enjoying a recent resurgence. See Nicholas Humphrey, 'Great Expectations: The Evolutionary Psychology of Faith Healing and the Placebo Effect', in *The Mind Made Flesh: Essays from the Frontier of Evolution and Psychology* (Oxford: Oxford University Press, 2002), pp. 255–85.
18. Stuart Cumberland, 'Illusionary and Fraudulent Aspects of Spiritualism', *Journal of Mental Science* (1881–82), pp. 280–87, 628–35; idem, 'A Thought Reader's Experiences', *Nineteenth Century,* 20, pp. 689; Roger Luckhurst, 'Passages in the Invention of the Psyche: Mind-Reading in London, 1881–84', in Roger Luckhurst and Josephine McDonagh (eds), *Transactions and Encounters: Science in Victorian Culture* (Manchester: Manchester University Press, 2002), pp. 117–50; Roger Luckhurst, *The Invention of Telepathy, 1870–1901* (Oxford: Oxford University Press, 2002).
19. Mark S. Micale, *Approaching Hysteria: Disease and its Interpretations* (Princeton: Princeton University Press, 1995).

20. Christopher Goetz, Michel Bonduelle, and Toby Gelfand, *Charcot: Constructing Neurology* (Oxford: Oxford University Press, 1995), p. 53, 241f; Chertok and Stengers, *Critique of Psychoanalytic Reason*, pp. 36–38.
21. Henri F. Ellenberger, *Discovery of the Unconscious* (1970), 171ff; Crabtree, *Mesmer*, pp. 164–68.
22. On Babinski's rejection of Charcot, see Joseph Babinksi, 'Definition de l'hysterie', *Revue Neurologie* (1901), 9, pp. 1074–80; idem, 'Démemberment de l'hystérie traditionelle: pithiatisme', *Semaine Médicale* (6 January 1909), pp. 66–7; Goetz, Bonduelle, and Gelfand, *Charcot*, pp. 321–2; Chertok and Stengers, *Critique of Psychoanalytic Reason*, pp. 230–33. On Hurst (1879–1944), see Alex Sakula, 'Sir Arthur Hurst: Master of Medicine', *Journal of Medical Biography* (1999), 7, pp. 125–29; Arthur Hurst, *A Twentieth Century Physician* (London: Edward Arnold & Co., 1949), especially chapter 6; Shephard, *War of Nerves*, pp. 78–80.
23. Arthur F. Hurst, 'Nerves and the Men', *Reveille* (1918), 2, pp. 260–8; idem, *Medical Diseases of War* (London: Edward Arnold, 1918).
24. Joseph Babinski and J. Froment, *Hysteria or Pithiatism, and Reflex Nervous Disorders in the Neurology of War*, trans. J.D. Rolleston, preface by E. Farquhar Buzzard (London: University of London Press Ltd., 1918), p. 223, 224ff; Marc Roudebush, 'A Battle of Nerves: Hysteria and its Treatments in France during World War One', in M. Micale and P. Lerner (eds), *Traumatic Pasts*, p. 262; Hurst, *Medical Diseases of War*.
25. Thomas Arthur Ross, *Common Neuroses* (London: Edward Arnold, 1923), p. 131. On Ross (1875–1941), see Richard R. Trail, *Lives of the Fellows of the Royal College of Physicians of London*, Munk's roll(London: Royal College of Physicians, 1968), vol. 5, 359–60.
26. Millais Culpin, *Psychoneuroses of War and Peace* (Cambridge: Cambridge University Press, 1920), pp. 27–8; idem, *The Nervous Patient* (London: H. K. Lewis, 1924), pp. 74–5. On Culpin see Frances Millais MacKeith, 'Culpin, Millais (1874–1952)', *Oxford Dictionary of National Biography* (Oxford: Oxford University Press, 2004), [http://www.oxforddnb.com/view/article/51592, accessed 10 Oct. 2005].
27. Frederick Lucien Golla, 'The Objective Study of Neurosis', *Lancet* (1921), 198, p. 376. On Golla see Rhodri Hayward, 'Golla, Frederick Lucien (1877–1968)', *Oxford Dictionary of National Biography* (Oxford: Oxford University Press, 2004), [http://www.oxforddnb.com/view/article/71668, accessed 10 Oct 2005]
28. Edmund Gurney, F. W. H. Myers and W. F. Barrett, 'Second Report on Thought Transference', *Proceedings of the Society for Psychical Research* (1882–83), 1, p. 70. For an overview of the Creery sisters, see Janet Oppenheim, *The Other World* (Cambridge: Cambridge University Press, 1991), pp. 358–61; Trevor Hall, *The Strange Case of Edmund Gurney* (London: G. Duckworth, 1964), pp. 55–63.
29. Bernard Hart, 'The Conception of the Subconscious', *Journal of Abnormal Psychology* (1910), 4, pp. 351–71; Mikkel Borch-Jacobson, *The Freudian Subject* (Ithaca, NY: Cornell University Press, 1988).
30. Armstrong Jones, 'Ministry of Healing: Psychological and Psychic', *Contemporary Review* (August 1926), p. 194. See also: Robert Armstrong Jones, 'Functional Nervous Disease', *Lancet* (1916), 187, pp. 210–11. On Armstrong Jones see W. J. Bishop, 'Jones, Sir Robert Armstrong (1857–1943)', revised

Nick Hervey, *Oxford Dictionary of National Biography* (Oxford: Oxford University Press, 2004), [http://www.oxforddnb.com/view/article/30453, accessed 10 Oct. 2005].

31. 'Listerian Lecture: Kings College Hospital, 1937', MS in Bethlem Royal Hospital Archive (BRHA) Series 1: Box EM2. For background see Rhodri Hayward, 'Making Psychiatry English: The Maudsley Hospital and the Munich Model', in Volker Roelcke and Paul Weindling (eds), *Inspiration, Co-operation, Migration: British–American–German Relations in Psychiatry, 1870–1945* (Rochester: University of Rochester Press, forthcoming 2006), p. 47.
32. Michael Neve and Trevor Turner, 'What the Doctor Thought and Did: Sir James Crichton-Browne (1840–1938)', *Medical History* (1995), 39, pp. 399–432; Trevor Turner, 'James Crichton-Browne and the Anti-Psychoanalysts', in Hugh Freeman and German Berrios (eds), *150 Years of British Psychiatry:* Volume II: *The Aftermath* (London: Athlone, 1996), pp. 144–55.
33. Cf. Golla, 'Objective Study'; idem, 'The Objective Study of Neurosis Pt. II', *Lancet* (1921), 198, pp. 215–221; idem, 'The Objective Study of Neurosis Pt. III', *Lancet* (1921), 198, pp. 265–70; idem, 'The Objective Study of Neurosis Pt. IV', *Lancet* (1921), 198, pp. 373–79; idem, 'The Organic Basis of the Hysterical Syndrome', *Proceedings of the Royal Society of Medicine (Section of Psychiatry)* (1923), 16, pp. 1–12.
34. Golla, 'Objective study of neurosis Part IV', p. 377.
35. For an overview, see Otniel E. Dror, 'The Scientific Image of Emotion: Experience and Technologies of Inscription', *Configurations* (1999), 7, pp. 355–401; idem, 'The Affect of Experiment: The Turn to Emotions on Anglo-American Physiology, 1900–1940'. *Isis* (1999), 90, pp. 205–237; idem, 'Techniques of the Brain and the Paradox of Emotions, 1880–1930', *Science in Context* (2001), 14, pp. 643–60. Geoff Bunn, 'The Lie Detector, Wonder Woman and Liberty: The Life and Work of William Moulton Marston', *History of the Human Sciences* (1999), 10, pp. 91–119; Rhodri Hayward, 'The Tortoise and the Love-Machine: Grey Walter and the Politics of Electro-encephalography', *Science in Context* (2001), 14, pp. 615–41.
36. 'The Betraying Light', *Daily Express* (11 Feb. 1920), commenting on Golla's colleague, A. D. Waller's experiments. See Augustus D. Waller, 'The Measurement of Human Emotion and of its Voluntary Control', *Proceedings of the Royal Society of Medicine* (1919–20), 13, pp. 41–56; James Barr, 'Measurement of Emotion', *British Medical Journal* (1920), 1, p. 381.
37. Roger Cooter, 'Malingering in Modernity: Psychological Scripts and Adversarial Encounters during the First World War' in Roger Cooter, Mark Harrison, and Steve Sturdy (eds), *War, Medicine and Modernity* (Stroud: Sutton Publishing, 1999), pp. 125–48.
38. For example, Robert Brudenell Carter, *On the Pathology and Treatment of Hysteria* (London: Churchill, 1853).
39. On Sinclair, see Suzanne Raitt, *May Sinclair: A Modern Victorian* (Oxford: Oxford University Press, 2000).
40. *The Helpmate* (London: Archibald Constable & Co, 1907). Cf. also the figures of Priscilla Heaven and Robin Lethbridge in Sinclair's final novel, *The Life and Death of Harriet Frean* (New York: The Modern Library, 2003 [1922]).
41. Q.v. Charles Rycroft, 'Gain, Primary and Secondary (Paranoic and Epinosic)', in *A Critical Dictionary of Psychoanalysis* (London: Penguin Books, 1972), p. 56.

42. Robert Hutchison, *Lectures on Dyspepsia*, 2nd ed. (London: Edward Arnold, 1927); Robert Dick Gillespie, *Hypochondria*, Psyche Miniatures Medical Series no. 12 (London: Kegan Paul Trench Trübner, 1929), p. 44; William James O'Donovan, *Dermatological Neuroses* (London: Kegan, Paul, Trench, Trübner, 1927), p. 171.
43. Walter Langdon-Brown, Robert George Macdonald Ladell, Frank Gray and Francis Graham Crookshank (eds), *The Place of Psychology in the Medical Curriculum* (London: C. W. Daniel, 1936); Hugh Crichton-Miller, 'Psychotherapeutic Clinics in Fact and Fancy', *British Medical Journal* (1935), 1, p. 1205.
44. Georg Groddeck, 'Psychical Treatment of Organic Disease', *British Journal of Medical Psychology* (1929), 9, pp. 179–186; Alfred Adler, *Problems of Neurosis: A Book of Case-Histories* with a prefatory essay by Francis Graham Crookshank; edited by Philippe Mairet (London: Kegan Paul, Trench, Trübner, 1929); idem, *The Case of Mrs A: The Diagnosis of Life-Style* Individual Psychology Medical Pamphlets, 1 (London: C. W. Daniel, 1931).
45. On Francis Graham Crookshank (1873–1933) see G.H. Brown, *Lives of the Fellows of the Royal College of Physicians of London, 1826–1925*, Munk's roll, vol. 4 (London: Royal College of Physicians, 1955), pp. 567–68; *British Medical Journal* (1933), 2, p. 848; Phyllis Bottome, *Alfred Adler: Apostle of Freedom*, 2nd ed., (London: Faber and Faber, 1953), pp. 212–13.
46. Luisa Passerini, *Love in Europe, Europe in Love: Imagination and Politics between the Wars* (London: Tauris, 1999); Andrew Rigby, *Initiation and Initiative: An Exploration of the Life and Ideas of Dmitrije Mitrinovic*, East European Monographs no. 164 (Boulder: EEM, 1984).
47. Francis Graham Crookshank, *The Mongol in our Midst: A Study of Man and his Three Faces*, 3rd ed., (London: Kegan Paul, Trench Trubner and Co., 1931).
48. Crookshank, *Mongol*, pp. 105, 462.
49. Chris Lawrence, 'A Tale of Two Sciences: Bedside and Bench in Twentieth-Century Britain', *Medical History* (1999), 43, pp. 421–49.
50. Adler, *Problems of Neurosis*, pp. vii–xxxvii.
51. Alfred Adler, *Study of Organ Inferiority and its Psychical Compensation* (New York: Nervous and Mental Diseases Publishing, 1917 [1907]); Francis Graham Crookshank, 'Organ States and Emotional Correlatives', *Psyche* (1930), 10, pp. 65–79.
52. Francis Graham Crookshank, *Migraine and Other Common Neuroses: A Psychological Study*, Psyche miniatures, Medical series 1 (London: K. Paul, Trench, Trübner, 1926), p. 40.
53. Walter Langdon Brown, 'Adler's Contribution to General Medicine', p. 50.
54. Francis Graham Crookshank, 'Psychological Interest in General Practice', *British Medical Journal* (1932), 1, pp. 599–604. See also idem, 'Organ Jargon', *British Journal of Medical Psychology* (1930), 10, pp. 295–311; idem, 'Organ Inferiorities' in Langdon Brown, Macdonald Ladell, Gray and Crookshank (eds), *The Place of Psychology;* idem, 'Organ States and Emotional Correlatives', *Psyche* (1930), 10, pp. 65–79.
55. William Samuel Inman, *Styes and Wedding Rings* (Edinburgh: Churchill Livingstone, 1973); idem, 'Eye Symptoms' in Millais Culpin (ed.), *Nervous Patient* (London: H. K. Lewis, 1924), Chapter 16; idem, 'The Symbolic

Significance of Glass and its Relation to Diseases of the Eye', *British Journal of Medical Psychology* (1939), 18, pp. 122–40.

56. Henry Yellowlees, *Manual of Psychotherapy for Practitioners and Students* (London: A. & C. Black, 1923), 106, pp. 181–82; idem, *Frames of Mind*, (London: William Kimber, 1957). On Yellowlees (1888–1971), see Gordon Ethelbert Ward Wolstenholme (ed.), *Lives of the Fellows of the Royal College of Physicians of London,* Munk's roll; 6 (London: Royal College of Physicians, 1982), pp. 481–82.
57. Ross, *Common Neuroses*, pp. 134–35.
58. Margiad Evans, *The Wooden Doctor* (Oxford: Blackwell, 1933).
59. Henry Brackenbury, *Patient and Doctor* (London: Hodder & Stoughton, 1935), p. 74; Hugh Crichton-Miller, 'Primitive Man and Modern Patient', *British Medical Journal* (1932), 2, p. 431.
60. William Halse Rivers, *Mind and Medicine: A Lecture*, reprinted from *Bull. John Rylands Library*, (1919), 5 (London: Longmans Green & Co., 1923); idem, 'Psychology and the War', *Instinct and the Unconscious*, (Cambridge: Cambridge University Press, 1922), Appendix 8, pp. 257–58.
61. M.O. Raven, 'Medicine as an Art', *Lancet* (1932), 220, pp. 602–3.
62. Maurice Cassidy, 'Doctor and Patient', *Lancet* (1938), 231, pp. 175.
63. Harry Bannister, *Psychology and health* (Cambridge: Cambridge University Press, 1935) p. 110.
64. 'The Importance of Suggestion', *Lancet* (1936), 228, p. 1432.
65. Walter Langdon Brown, 'New Pains for Old', in idem, *Thus we Are Men* (London: K. Paul, Trench, Trubner and Co. Ltd, 1938); idem, 'Individual Psychology and the Sympathetic Mechanism', *British Medical Journal* (1931), 2, p. 753.
66. Douglas-Fairhurst, *Victorian Afterlives*, pp. 87ff. Jonathan Arac, *Commissioned Spirits: The Shaping of Social Motion in Dickens, Carlyle, Melville, and Hawthorne*, (New Brunswick, NJ: Rutgers University Press, 1987).
67. Francis Graham Crookshank, *Epidemiological Essays* (London: Kegan Paul, 930), pp. 14–28; idem, *Influenza* (London: William Heinemann (Medical Books), 1922). For discussion, see Andrew J. Mendelsohn, 'From Eradication to Equilibrium: How Epidemics became Complex after World War One', in Christopher Lawrence and George Weisz (eds), *Greater than the Parts: Holism in Biomedicine, 1920–1950* (Cambridge: Cambridge University Press, 1998), p. 311.
68. Francis Graham Crookshank, 'Types of Personality with Special Reference to Individual Psychology', *Lancet* (1930), 215, pp. 546–48.
69. Francis Graham Crookshank, 'Supplement II: The Importance of the Theory of Signs and the Critique of Language in the Study of Medicine', in Charles Kay Ogden and Ivor Armstrong Richards (eds), *The Meaning of Meaning: A Study of the Influence of Language upon Thought and of the Science of Symbolism* (London: International Library of Philosophy, Psychology and Scientific Method, 1923), pp. 511–37.
70. 'The Name and the Word: Neo-Hippocratism and Language in Interwar Britain', in David Cantor (ed.), *Reinventing Hippocrates* (Aldershot: Ashgate, 2002), pp. 280–301.
71. Edward Mapother, 'Tough or Tender. A Plea for Nominalism in Psychiatry', Presidential address of the *Proceedings of the Royal Society of Medicine* (1934), 27, pp. 1687–712.

72. Robert Hutchison, 'Hypochondriasis: Individual, Vicarious and Communal' *British Medical Journal* (1934), p. 365. This provoked a large correspondence in this volume, pp. 505, 548, 643, 731, 776, Annotation, p. 541.
73. Gillespie, *Hypochondria*, p. 94.
74. Robert Dick Gillespie, *The Mind in Daily Life* (London: Methuen, 1933).
75. 'Morrison Lectures', *Edinburgh Medical Journal* (1935), 42, p. 445; Cf. Langdon-Brown, Macdonald Ladell, Gray and CrookShank (eds), *The Place of Psychology*.
76. Christopher Lawrence, 'Edward Jenner's Jockey boots and the Great Tradition in English Medicine, 1918–39', in Christopher Lawrence and Anna-K. Mayer (eds), *Regenerating England: Science, Medicine and Culture in Inter–War Britain* (Amsterdam: Rodopi, 2000), pp. 45–66.
77. Cyril Donnison, *Civilization and Disease* (London: Balliere, Tindall and Cox, 1937); Christopher Lawrence, 'Still Incommunicable: Clinical Holists and Medical Knowledge in Inter–war Britain', in Christopher Lawrence and George Weisz (eds), *Greater than the Parts: Holism in Biomedicine, 1920–50* (Cambridge: Cambridge University Press, 1998), pp. 94–111; Charles E. Rosenberg, 'Pathologies of Progress: The Idea of Civilization as Risk', *Bulletin of the History of Medicine* (1998), 72, pp. 714–30.
78. Galatians 4:19; William Sanday, '*Outlines of the Life of Christ* (Edinburgh: T. & T. Clark, 1906), p. 228. There was a considerable literature which argued for the formation of character through the imitation of Christ: William Straton Bruce, *The Formation of Christian Character* (Edinburgh: T. & T. Clark, 1902), Chapter 5, pp. 87–93; Richard William Church, *The Discipline of Christian Character* (London: Macmillan, 1885), pp. 102–13, 128–30; John Richardson Illingworth, *Christian Character* (London: Macmillan, 1904), pp. 33–8; James Russell Miller, *The Building of Character* (London: Sunday School Union, n.d.), Chapter 3.
79. Arthur Taylor Schofield, *Unconscious Therapeutics* (London: J. & A. Churchill, 1904), Chapter 4; idem, *Springs of Character* (London: Hodder & Stoughton, 1905), 143ff; Janet Oppenheim, *Shattered Nerves: Doctors, Patients and Depression in Victorian England* (New York; Oxford: Oxford University Press 1991), pp. 139–40.
80. Dyce Duckworth, *Harveian Oration* (London: Longmans, 1898).
81. Francis Graham Crookshank, *Diagnosis: and Spiritual Healing*, Psyche miniatures, Medical series, 6 (London: Kegan Paul, Trench, Trübner, 1927), p. 95; Charles Maurice Bevan Brown, 'Psychological Schools a Plea for Correlation' in Charles Maurice Bevan Brown, G. E. S. Ward and Francis Graham Crookshank (eds), *Individual Psychology Theory and Practice*, I.P. Pamphlet, 15 (London: C. W. Daniel, 1936), pp. 9–42 especially p. 20.
82. Adrian C. Laing, *R. D. Laing: A Biography* (London: Peter Owen, 1994), pp. 64, 87–9; John Clay, *R. D. Laing: A Divided Self* (London: Hodder & Stoughton, 1996), p. 77; James Webb, *Occult Establishment* (La Salle, Ill.: Open Court Publishing, 1976), p. 476.
83. Eric Graham Howe, 'Motives and Mechanisms of the Mind. XII: Psychopathology in Relation to Treatment', *Lancet* (1931), 220, pp. 714–21; idem, *Motives and Mechanisms of the Mind: An Introduction to Psychopathology and Applied Psychology*, Post-graduate Medical Series, 1 (London: The Lancet, 1931), Chapter 12.

84. Phyllis Bottome, *Private Worlds* (London: Bodley Head, 1934). See idem, *Alfred Adler: Apostle of Freedom* (London: Faber and Faber, 1957 [1939]).
85. *Private Worlds*, dir. Gregory la Cava (Paramount, 1935).
86. W. R. Houston, 'The Doctor Himself as a Therapeutic Agent', *Annals of Internal Medicine* (1938), 11, pp. 1416–25, Cf. M. Foucault, 'About the Beginning of the Hermeneutics of the Self', *Political Theory* (1993), 21, pp. 203–23; Luther H. Martin, Huck Gutman and Patrick H. Hutton (eds), *Technologies of the Self: A Seminar with Michel Foucault* (London: Tavistock, 1988).
87. David Armstrong, 'The Doctor–Patient Relationship, 1930–80', in Wright and Treacher (eds), *The Problem of Medical Knowledge*, pp. 109–22.
88. Thomas Osborne, 'Mobilizing Psychoanalysis: Michael Balint and the General Practitioners', *Social Studies of Science* (1993), 23, pp. 175–200; Michael Marinker, 'What is Wrong' and 'How We Know It': Changing Concepts of Illness in General Practice' in Irvine Loudon, John Horder and Charles Webster (eds), *General Practice under the National Health Service* (London: Clarendon Press, 1999), pp. 68–9.
89. M. J. F. Courtenay, 'The Place of Balint Work in Medicine: Looking Back – Looking Forward', *Journal of the Balint Society* (1984), 12, pp. 8–12.
90. Michael Balint, *The Doctor, his Patient and the Illness*, (London: Pitman, 1956), pp. 222ff.
91. Ibid., p. 224.
92. Ibid., p. 235.
93. Thomas Osborne, 'Power and Persons: On Ethical Stylisation and Person-Centred Medicine', *Sociology of Health and Illness* (1994), 16, pp. 515–35.
94. Barry Barnes, 'Social Life as Bootstrapped Induction', *Sociology* (1983), 17, pp. 524–45; Ian Hacking, 'The Looping Effects of Human Kinds', in Dan Sperber, David Premack, and Ann J. Premack (eds), *Causal Cognition: A Multidisciplinary Approach*, (Oxford: Clarendon Press, 1994), pp. 351–94. On emotion, see Rom Harré (ed.), *The Social Construction of Emotion*, (Oxford: Blackwell, 1986); Kurt Danziger, *Naming the Mind: How Psychology found its Language* (London: Sage Publications, 1997). On expectation, Ludwig Wittgenstein, *Philosophical Investigations* (Oxford: Basil Blackwell, 1953), pp. 444–53.
95. For an overview, see Dylan Evans, *Placebo: Mind over Matter in Modern Medicine* (London: Harper Collins, 2004). For thoughts on the historiographical importance of the placebo effect, see David Harley, 'Rhetoric and the Social Construction of Illness and Healing', *Social History of Medicine* (1999), 12, pp. 407–35. For the impact of new illness categories on lived experience and social organisation, see Allan Young, *The Harmony of Illusions: Inventing Post-Traumatic Stress Disorder*, (Princeton, NJ: Princeton University Press, 1995).

8

Diagnosing with Feeling: The Clinical Assessment of Schizophrenia in Early Twentieth-Century European Psychiatry[1]

Susan Lanzoni

> 'To acknowledge someone as a sane person is to establish a reciprocal relation to him.' Michael Polanyi.[2]

Schizophrenia epitomized the incomprehensible for many European psychiatrists of the early twentieth century.[3] Eugen Bleuler, the Swiss psychiatrist who brought the term schizophrenia into common use beginning in 1911, described the associations schizophrenics made as unexpected and bizarre.[4] The disorder was seen to defy the clinician's capacity to decipher the patient's motives or intentions, making it impossible to empathize with the patient. Bleuler differentiated the schizophrenic from the neurasthenic by the absence of an affective connection (Gefühlsrapport) between patient and doctor, a connection that was easily forged with the neurasthenic and even manic-depressive.[5]

Coupled with the blocking of a rapport, clinicians often reported strange feelings or experiences when in the presence of a schizophrenic. In the 1920s, phenomenological psychiatrists such as Ludwig Binswanger and Eugène Minkowski made what they called a 'feeling diagnosis' (Gefühlsdiagnose) by attending to the feelings of alienation aroused in the clinician by the schizophrenic patient. In 1941, the Dutch psychiatrist Henricus C. Rümke dubbed the strange feeling a clinician experienced while meeting with a schizophrenic, the *'praecox* feeling' (Praecoxgefühl). Rümke argued that the clinician could attain greater diagnostic reliability by relying on this feeling, which could differentiate schizophrenia from confusional manias, schizoid psychopathies and degenerative psychoses. Also in this period, the Swiss psychiatrist Jakob Wyrsch argued that an experienced clinician could employ a holistic, intuitive perception to perceive the essence of a schizophrenic disturbance. For these practitioners, the registered emotion and experience of the psychiatrist served as important instruments of diagnosis.

This article tracks these epistemologies of clinical experience and emotion as prevalent but contested modes of psychiatric knowledge and expertise in early to mid-twentieth-century Continental Europe. Recent literature on the cultural and scientific history of the emotions and sentiment has contested the view of the emotions as merely private and non-rational experiences, and has rightly offered more complicated accounts of the emotions as culturally situated and scientifically shaped.[6] Yet little had been written on emotions and experience as an instrument of knowledge in scientific and medical settings.[7] For European psychiatrists loosely grouped together by an interest in phenomenological or descriptive methods, the clinician functioned as a tool for psychiatric diagnosis by responding as a tuned emotional instrument to the mental states of his or her patients. These practices extended beyond clinical judgment to a technology of the self for clinicians, who argued that their own responsive feeling to the patient enabled a more definitive grasp of the essence of the disorder than an enumeration of symptoms.

These psychiatrists were part of an interpretative tradition in the human sciences that made use of understanding, empathy and other psychological means for diagnosis.[8] As advanced by the German philosopher Wilhelm Dilthey, and picked up by a range of philosopher/psychologists, the methods of the human sciences included an empathic or emotional projection into the lives of others, a tool that was of use in the historical, literary and psychological sciences.[9] Armed with such methods, psychiatrists sought to re-vitalize the importance of clinical insights at a time when physiological and laboratory models had become more prominent. Most strikingly, they embraced a participatory or interactive epistemology, in which empathic methods and their failure could reveal the nature of patients' disorders within the matrix of the doctor–patient encounter.

For these clinicians, examining the ways others were perceived in everyday encounters formed an important clue for ascertaining the essential nature of the schizophrenic disturbance. These clinicians agreed that the schizophrenic was marked by an inability to engage in common social exchange with others, by an unmooring from his or her world, or a failure to connect with reality. Thus, the schizophrenic was viewed as standing outside of the human community. As the Polish psychiatrist Minkowski reflected, due to the loss of the ability to communicate with others, the schizophrenic was no longer one's fellow creature (semblable).

In an encounter with such a patient, then, one had to conclude that he was a madman (aliéné).[10] The knowledge gained in the clinical encounter thus led to an ontology of schizophrenia that emphasized the ways in which the schizophrenic embodied the quintessential 'other'.

The emphasis on the clinician's emotional experience in diagnosis, however, challenged some deeply engrained notions of what counted as scientific knowledge. At a symposium on the taxonomy of mental disorders in 1959, the logical empiricist Carl Hempel argued that a feeling response was not operationally specified nor publicly replicable.[11] Nonetheless, it was also true that little in psychiatric diagnostics fulfilled these requirements at this time, as quickly became evident at the symposium. The lack of rapport with schizophrenics continued to be discussed by clinicians at mid-century and afterwards, and remained a tacit aspect of clinical diagnostic practice resistant to systematization.[12] This chapter explores these embedded epistemologies of psychiatric diagnostics at the borders of psychiatric science, clinical practice and everyday experience.

Clinical experience in European psychiatry

Asylum and clinical psychiatrists of various stripes in early twentieth-century Continental Europe argued for the importance of descriptive methods in clinical practice, especially in the face of the growing popularity of physiological and experimental approaches. Despite the initial excitement over the discovery that general paralysis was of syphilitic origin, there was increasing doubt regarding the ability to find direct anatomical or physiological correlates of psychiatric disturbances. For example, the findings of prominent psychiatrists such as Alfred Hoche and Karl Bonhoeffer had demonstrated that there was no simple one-to-one correlation between organic damage and psychological response.[13] In the Zurich school in Switzerland, psychiatrists brought renewed focus to the clinical realm and psychosocial aspects of disorders, in part prompted by the lack of national funding for laboratory research.[14] Eugen Bleuler and a number of his students looked to developments in psychology for inspiration, a search that sometimes extended to philosophy and phenomenology. In addition to the Zurich school, psychopathological approaches that achieved prominence in the early years of the twentieth century included Emil Kraepelin's descriptive tradition and the works of Sigmund Freud and Karl Jaspers.[15]

It was in this period that a phenomenological orientation in European psychiatry began to gain influence. Phenomenological psychiatrists argued that mental disorders should be given careful description without the aid of theoretical presuppositions, and they gave a great deal of attention to patients' own reports of illness experiences.[16] Karl Jaspers was one of the founders of this approach in psychiatry and drew attention to patient reports as a valuable source of psychopathological information. In his highly influential *General Psychopathology* of 1913, he documented the myriad forms of psychopathological experience with numerous first-person reports, culled from his own and other psychiatrists' clinical practices. He argued that this evidence provided the best scientific data for analysis:

> Our chief help in all this comes from the patients' own self-descriptions, which can be evoked and tested out in the course of personal conversation. From this we get our best-defined and clearest data ... An experience is best described by the person who has undergone it ... Psychotics' self-descriptions are not only unique but yield reliable results and through them we have discovered many of our basic concepts. If we compare what patients say we find much that is similar.[17]

A close analysis of the thoughts and language of patients was also stressed by Eugen Bleuler, director of the Burghölzli hospital in Zürich from 1898 to 1927. Bleuler had defined schizophrenia as a splitting apart of mental functions in 1911, a term that replaced the Kraepelinian 'dementia praecox', which had instead stressed the deteriorating course of the illness. Bleuler's outline of the fundamental simple symptoms—disorders of association, affectivity and ambivalence—was preeminently a psychological characterization of the disorder, based in large part on patients' language production. For both Bleuler and for Carl Jung, his collaborator at the Burghölzli, affect played an important role in their psychological approach to schizophrenia, particularly in the determination of the psychological complex as affectively motivated.[18] Bleuler's psychological approach was inspired by the relatively new Freudian psychoanalysis and in turn it fostered developments in phenomenological psychiatry. Indeed, many who had trained with Bleuler went on to develop phenomenological approaches including Ludwig Binswanger, Eugène Minkowski, and Henricus Cornelius Rümke.

In this tradition, diagnosing as well as understanding pathological mental processes were centered in the domain of clinical practice. Clinical practice often involved close contact with patients, with duties extending beyond the regular workday. Psychiatrists at the Burghölzli were known for working long hours, in intimate contact with patients.

Bleuler himself had a reputation for working closely with patients, speaking the local dialect, and trying to establish an emotional rapport with them.[19] As director, Binswanger lived on the premises of the Bellevue asylum, and Minkowski, in at least one instance, lived with his delusional patient. In some cases, the attention lavished on patients' own reports of illness was a function of their class status; for instance, the clientele of Binswanger's elite Bellevue asylum were among the wealthiest and most established in Europe. In such private settings, upper-class, often well-educated patients provided reflective reports of their own experience, and were listened to by their physicians.[20] But it was also true that in less elite clinics, for instance at the University of Utrecht, where Rümke saw patients, attention was also paid to the subjective meanings of patients' utterances. As Rümke declared in a series of lectures given in Denmark in 1953, the clinic was the center of psychiatry, despite or because of the fact that a science of differential diagnosis had still not been achieved in psychiatry.[21] For these practitioners, the clinical sphere was thus the privileged locus for a feeling-based and intuitive understanding of mental disorders which at times, could be superior to more systematic forms of knowledge.

Diagnosing with affect

Jaspers's pioneering phenomenological approach strove to make sense of mental disorder through patient reports of experience, yet it had its limits with schizophrenic patients. According to Jaspers, the clinician could not empathize with the schizophrenic's mental processes, and thus, unlike those with affective disturbances, the schizophrenic was beyond understanding.[22] This lack of empathy was similar to the ways in which a layperson could clearly demarcate the difference between the genuinely crazy (verrückt), and those with less severe disturbances. In Jaspers's view, it was this common-sense epistemology, borrowed from models of everyday interaction, which was of value to psychiatric diagnostics. However, it needed a clearer and more precise articulation.

One who attempted to do just that was Ludwig Binswanger, the Swiss director of the private Bellevue asylum. Binswanger was a philosophical autodidact, and had trained with Eugen Bleuler and Carl Jung at the Burghölzli Hospital in Zürich. He had been one of the first psychiatrists to introduce psychoanalysis into an asylum context. However, he shifted from a psychoanalytic framework to a phenomenological one in the early 1920s, and a decade later, published his first studies in existential anthropology, later to culminate in his method of existential analysis

(Daseinsanalyse). *Daseinsanalyse* was a hybrid method employing existential philosophical concepts in the analysis of psychopathological cases of mania and schizophrenia.

In a 1924 lecture on the relationship of psychology to psychiatry, Binswanger reported that the psychiatrist could sometimes base a diagnosis on the feelings evoked in the presence of the patient (Gefühldiagnose).[23] He may have adopted this term from Bleuler, who had referred to a *Gefühlsdiagnose* in 1906 as a form of intellectual feeling, also described as an unclear perception, conclusion or cognition.[24] Binswanger described a feeling-diagnosis as a kind of an 'emotional perception':

> We diagnose in this case actually not according to, but *with* our feeling, that is by means of that manner of perception, which, aside from the name, has not the slightest in common with 'feeling' in the sense of sensory or emotional feelings.[25]

He described this feeling as one of being rebounded back into oneself, which occurred when the appearance and manner of a schizophrenic entering a room so astounded the clinician. More pointedly, it was an experience of a barrier, or an inability to extend oneself in an understanding manner toward the patient. A feeling-diagnosis could result from the perception of the schizophrenic's characteristic glance or way of looking; it could result from extended time spent with the patient, during which one noticed a lack of emotional rapport; or it could result from a single encounter. Binswanger differentiated this reaction from a mere attraction or repulsion and explained:

> [A] schizophrenic can be very likeable to me as a person, and yet I rebound (zurückprallen) internally, I always experience a barrier of my own inner unity with him, in the perception of his person as a schizophrenic.[26]

This process was not just a mysterious reaction, however, as one could be trained to register and utilize these perceptions for further judgments and conclusions. Indeed, just as one could precisely and critically analyze one's external perceptions of another's physical characteristics, according to Binswanger, so one could further investigate the perception of another's mental life. This rebounding feeling was thus rooted in an actual perception of the other and was a reliable indication of their mental state.

Another psychiatrist who found merit in the *Gefühlsdiagnose* was Binswanger's colleague, Eugène Minkowski, a Polish-born psychiatrist who had also trained with Bleuler at the Burghölzli. Minkowski had obtained his medical degree in Munich, and before the First World War began attending lectures and reading the phenomenological philosophies of Max Scheler and Henri Bergson. After volunteering for the French army he settled in France, and worked as a psychiatrist at the Saint-Anne Hospital in Paris.[27] In a 1924 letter exchange with Binswanger, Minkowski expressed interest in the *Gefühlsdiagnose*, but wondered if there was a more neutral term for it. He was afraid that the rationalistic nature of French psychiatry would reject the term. Binswanger responded that one could also label it an intuitive diagnosis as carried out by a gifted physician, but more pointedly it could be called an inner perception, in Max Scheler's philosophical terminology.[28] What Binswanger meant by this was that one could evaluate the perception of another's psychological state (as one might say today, the perception of other minds) as a direct, unmediated and reliable form of perceptual knowledge. Scheler had designated the perception of others' minds an inner perception in contrast to an outer perception that attended to physical or sensory data. In searching for alternate terms for this phenomenon, Binswanger suggested using the term 'diagnostic intuitif' to characterize a doctor's intuition, in distinction to the direct perception of another's psychological state, which could be called: 'diagnostic direct, immédiate, or phénoménologique' or 'psychologique direct'. The latter terms emphasized the everyday access to another's inner state, a common process that was not restricted to the talented physician.

Minkowski expanded upon Binswanger's concept of a *Gefühlsdiagnose* in his 1926 book *La Schizophrénie*, by declaring that it was not only a feeling but also the psychiatrist's personality that could be an instrument of diagnosis. He redefined the diagnosis of feeling (le diagnostic par sentiment) as one that proceeded 'through penetration', which was similar to one's everyday interactions with others.[29] One did not come to understand another's personality by rationally cataloging his or her reactions, but by employing intuition so as to penetrate the other's personality. To do so one attended to one's own sensations, for instance, those of coldness or heat felt in the presence of the other. In a similar way, the psychiatrist was to affectively put him or herself in tune with the patient.[30] It was not enough to observe impassively or to adopt an analytic perspective as if looking under a microscope; rather one must 'put all of one's own personality into play, to confront the particular character, from the affective point of view, that emerged from their reactions as a whole'.[31] In doing so, one was not to pity the patient, but 'try to "feel" (sentir) with him and to see how he feels'.[32]

Minkowski's method of putting the psychiatrist's personality and feelings into play had already been put into practice in a case study he presented at a conference in Zürich hosted by the Swiss Society of Psychiatry in 1922. His talk was followed by remarks given by Binswanger on the uses of phenomenology in clinical psychiatry, and Bleuler also provided commentary during the session.[33] Minkowski's study was based on a delusional schizophrenic patient who believed that at the end of each day he would be physically destroyed by having all the waste in the world poured into his stomach. Minkowski had lived with the patient for two months as his personal physician, and he reported that this extended contact made it impossible to view the patient in a medical light 24 hours a day. The clinical encounter in this case went far beyond a cursory patient–doctor interaction, and presented an opportunity to delve into deeper levels of the patient's experience. It also fostered the constant comparison of the diseased psyche with his own, which Minkowski called the method of 'penetration':

> It was like two melodies being played simultaneously; although these two melodies are as dissonant as can be, a certain balance becomes established between the notes of the one and the other and lets us penetrate a little further into our patient's psyche.[34]

In delusional moments, the patient accused Minkowski of participating in the plot against him, but after such an episode, the patient would usually enter a depression, and try to re-establish contact with Minkowski in a relatively normal manner.

Minkowski concluded that the discordance between the patient's psyche and his own lay not in the content or even the existence of the delusion, but in the patient's experience of time. The temporal disturbance was that the patient faced the future with dread, with each day isolated from the previous one, and this disturbance formed the root of his delusion.

In his book on the nature of experienced time, *Le Temps Vécu*, published in 1933, Minkowski continued to stress the importance of the psychiatrist's feelings in the presence of a patient for the understanding of mental disorders.[35] He described a patient who had been a colonel in the Russian army who claimed that he was the genius behind the Allied offensive plan and thus suffered from delusions of grandeur and persecution. Minkowski tried to make a connection to the emotional elements of the patient's disorder, and in doing so, he speculated that a current of sympathy, or communion of ideas between himself and the patient could be created.[36] The patient resided at the clinic and Minkowski watched him each day while he was engaged in various tasks. One day, while he was listening to this patient recount his delusions, Minkowski felt a certain sentiment surge inside him, which he said was captured by the phrase, 'I know all about him' ('Je sais tout de lui').[37]

In Minkowski's analysis of the meaning of this phrase, he first considered that it might mean he had gained new knowledge of the patient, or alternately, that he had found the patient boring. He rejected these interpretations, however. He noted that the thought was accompanied by a certain malaise, or a feeling that signaled an impoverishment or a break in the habitual relations between people. It was therefore an unusual feeling, as he put it, 'In the presence of normal individuals we never experience a feeling of this order'.[38] In everyday social exchanges, it was impossible to have a feeling of the absolute knowledge of another, as one could always sense some aspect of the unknown in the other. In contrast, Minkowski's feeling of knowing all about this Russian soldier served as a sign of the patient's pathology. As Minkowski put it, the patient's life sources were exhausted, there was no depth to his life, and his persecutory, bizarre thoughts had taken over his being. The patient's psyche had lost its dynamic, animated nature, and was splayed out in a static assortment of objects or details, forming the elements of his delusion.

Minkowski called the static nature of the schizophrenic psyche a loss of vital contact with reality, a notion that he had drawn from Bleuler's depiction of autism as one of the complex symptoms of schizophrenia.[39] Vital contact with reality meant an interchange or interpenetration of one's own processes of becoming with one's environment and with others; it was a feeling of harmony, but one that was dynamic and active, not static and at rest.[40] Minkowski also called it a syntonic impulse, which was present in the manic-depressive, and opposed to the schizoid impulse. This dichotomy was drawn from the works of Ernst Kretschmer and Eugen Bleuler, and delineated universal human impulses that bridged both the pathological and the normal.[41] Syntonic or schizoid tendencies gave individual symptoms and the entire mental state a particular coloring or complexion.[42] A diagnosis was supposed to grasp the overall complexion of the patient's mental state in a synthetic way, not merely catalog symptoms. By assessing the patient as having syntonic or schizoid tendencies, one could access the essential, or what Minkowski called, the soul or psyche (l'âme) of the mad.[43]

Minkowski's emphasis on the holistic impression as a basis for diagnosis was also central to the work of his contemporary, the Swiss psychiatrist Jakob Wyrsch, the director of the psychiatric clinic Waldau-Bern, and Professor of Psychiatry at the University of Bern. He wrote on the schizophrenic personality, and was closely connected to developments in existential psychiatry, helping to name Binswanger's form of existential analysis *Daseinsanalyse* in 1942.[44] In 1946, Wyrsch wrote that because of his extensive clinical experience, he often received an impression (Eindruck), or a feeling (Gefühl) of schizophrenia from the patient.[45] It was a feeling that something was missing or lacking, a feeling of absence. This absence was a decisive factor in the diagnosis, and the enumeration of symptoms would generally lend support to this basic diagnostic feeling.

Although Wyrsch employed the language of feeling, and saw use in the term *Gefühlsdiagnose*, he emphasized that the impression received by the clinician was an actual recognition—it was not a guess, or a speculation, or a mere subjective emotion.[46] Even though one could not express it sufficiently in words, this recognition could become clear through the life history of the patient, or through a depiction of the course of the disease, but most reliably, was made manifest in the presence of the patient. As he put it, descriptions of a patient could never do justice to a perception or intuition (Anschauung), which emerged in extended contact with patients over the course of many years.[47] To see a patient in this way meant that one had to strip away psychopathological theories and a medical perspective to view the patient as a fellow human being (Mitmenschen). This meant to bring the naïve truths of the everyday perception of others into the realm of clinical practice.

The fact that the schizophrenic struck one as strange and different (andersartig) in a commonsense way (volkstümlichen Sinne) formed an important beginning to the diagnostic process. It was not the impossibility of empathy that was the decisive means for the intuitive diagnosis, according to Wyrsch, but a perception that the schizophrenic was unmoored to his or her world.[48] Being unmoored meant they were detached and disconnected from others, their environment, and even from the histories, traditions and customs in which they had matured. This disconnection was reminiscent of Minkowski's claim that the schizophrenic was not making vital contact with reality, and it was this trait both deemed a core deficit in schizophrenia.

The language of empathy and feeling was preeminent, however, in the work of the Dutch psychiatrist Henricus C. Rümke, who coined the term 'praecox feeling' in 1941 to describe the clinician's affective response to a schizophrenic patient.[49] Rümke had ties to the phenomenological tradition, had also studied with Bleuler at the Burghölzli, and penned a monograph on the feeling of happiness for both the normal and mentally impaired in 1924. He held a position in developmental psychology at the University of Utrecht beginning in 1933 in addition to his clinical responsibilities. He wrote extensively on the diagnosis of schizophrenia, expressing concern that there was a tendency to over-diagnose the disorder. One important difference between the genuine and the pseudo-schizophrenic was that the latter could be cured by psychotherapeutic or somatic therapies but the former could not. According to Rümke, those who reported recoveries to relatively new treatments like electroshock had been diagnosed with something other than schizophrenia. Genuine schizophrenics, in contrast, were resistant to treatment, and thus fit within the deteriorating course laid out by Kraepelin in his characterization of dementia praecox.

In 1941, Rümke argued that the praecox feeling allowed the clinician to define his patient's mental state as possessing a particular schizophrenic coloring, and this was a reliable diagnostic indicator. As Rümke put it, this 'feeling, induced in the clinician, is the final and most important guideline'.[50] Without this guideline, a mere enumeration of symptoms could miss what was essential in the diagnosis, and overlook cases of genuine schizophrenia. Although a host of symptoms could be detected in the schizophrenic, including a loss of contact with reality, loss of the experience of ego activity, mood disorder, unchaining of the archetypal world, thought disturbances and delusions, these were not in themselves adequate markers for the disorder. Indeed, symptoms like catatonia and delusions were not necessarily schizophrenic. In other cases, degenerative hysterias, or infantile autism, as described by Kanner, had been mistaken for schizophrenia. There were also cases of psychoses induced by drugs, which evoked something that could be mistaken for the praecox feeling in the examiner, and Rümke admitted that he had made this mistake at times.

For Rümke, the praecox feeling was something not present in an individual symptom, but was 'comprehended in all the categories of the conventional mental state examination'.[51] In line with Jaspers, it was a lack of empathy that the examiner noticed in himself, a recognition that 'it is impossible to establish contact with the patient's personality as a whole'.[52] The clinician felt a certain one-sidedness in the interaction, or that in some sense the patient could not be located in the exchange. This culminated in feelings of despair on the part of the clinician, usually resulting in futile attempts to make a connection with the patient. For Rümke, the clinician's feelings of one-sidedness stemmed from a deficit in the patient, which he identified as the loss of a directedness toward other people and the environment. This weakening of the 'rapprochement-instinct', which was common to all humans as social creatures, thus comprised the core schizophrenic deficit.

Yet, the praecox feeling was only one example of a host of different possible reactions on the part of the doctor to the patient. Indeed, Rümke argued that a variety of feelings aroused in the clinician were vital, but often overlooked clinical instruments:

> The doctor's internal attitude induced by the patient is a very sensitive diagnostic tool, and it would be helpful if we were more skilled in recognizing changes in our own internal attitude; it would certainly make us more self-confident in making diagnoses.[53]

Indeed, there was often a mirroring of the patient's mental state in that of the physician, for instance, a manic patient might actually cause mania in the doctor; or a neurotic patient could provoke a temporary neurosis in the doctor. For Rümke then, clinicians' feelings were not only important diagnostic tools, but also ones that had to be controlled and monitored throughout the diagnostic process.

The personality of the psychiatrist thus played an important role in diagnosis, and Rümke advised that among other traits, the psychiatrist was to possess a complex and differentiated personality. As he reflected, 'For how else will he be able to feel and test the subtleties of the problems of his patients, who are often richly variegated personalities themselves?'[54] Although he acknowledged that Freud had explored the nature of the doctor–patient interaction, Rümke concluded that more research needed to be done. In 1910, Freud had labeled the analyst's own attitudes in the treatment process a counter-transference reaction, a reaction that often formed an obstacle to treatment. However, counter-transference did not play a large role in psychoanalytic theorizing in the early part of the twentieth century, except in the work of the analyst Sandor Ferenczi in the 1920s, and then later in the writings of Michael and Alice Balint.[55] In addition, these analysts dealt with patients with less severe symptoms, as they did not treat schizophrenics. Rümke wanted to place more emphasis on the role of the therapist, as he saw the doctor–patient relationship as a meeting of two personalities, in which the doctor's personality could never be totally effaced. As he intoned, 'I am convinced that in the experience of the therapist a wealth of unexplored material is present, which could be made very useful to others'.[56]

Problems of psychiatric taxonomy

Despite the stated importance of the praecox feeling to many clinicians, its status as a scientific procedure was contested at an international Conference on Field Studies in the Mental Disorders held in 1959 by the American Psychopathological Association.[57] Psychiatrists gathered from across Europe, Argentina and Canada to discuss the inconsistencies in psychiatric diagnosis across different national contexts. Many were dissatisfied with the International Statistical Classification of Disease (ICD) developed by the World Health Organization (WHO) in 1948. The psychiatric section of this Standard Classification (Section V) had been rejected by most member countries including the United States, Italy, Sweden and Norway, which had all developed their own classification schemes.[58] The plethora of classification systems resulted in great difficulties conducting cross-cultural field psychiatric research and determining rates of disease in different locales.

In addition to the ICD, there were other classifications, including the first Diagnostic and Statistical Manual (DSM) developed by the American Psychiatric Association in 1952, and more informal schemes used by individual psychiatrists. Some were centered on etiology, others on symptomology. To many psychiatrists, the DSM had the advantage of separating out the organically caused mental disorders of the brain from all other groups of disorder. As the German psychiatrist Meyer intoned, the benefit of this approach reduced the personal equation in diagnosis, which was ardently desired. Indeed, he likened the contemporary situation in psychiatry to that of the 'general physician before the discovery of bacteria'.[59]

The logical empiricist Carl Hempel opened the conference with an influential paper adumbrating a philosophical foundation to the problems of psychiatric taxonomy. He argued that at the first stage, taxonomies of mental disorder were to rely on observation, but at later stages on systematic theories. As Hempel described it, the development of science reflected a movement from simple observations to theoretical and thus predictive capacities. In psychiatry, however, the current theoretical models were psychoanalytic, and Hempel doubted that these theories had uniform criteria of application. In his discussion, Hempel spoke of the praecox feeling as an example of a diagnostic procedure that could not be operationally specified, meaning that it did not have 'definite public criteria of application'.[60] Although Hempel argued that operationism did not always involve experimental or instrumental manipulation—direct observation could also count as an operation—the praecox feeling still did not qualify.[61] The problem was that the criterion for the praecox feeling was not separate from the perceiving observer. As Hempel put it, 'the occurrence of the specified outcome, the praecox feeling in regard to a given patient, is *not* independent of the examiner'.[62]

Yet it became clear in the course of discussions over the four-day conference that the reliability of diagnoses and nomenclatures, as well as the meanings of psychiatric classifications were often not independent of individual examiners. There were wide discrepancies in diagnostic patterns across countries, psychiatric schools, hospitals and even between individual practitioners. Commentators pointed not only to conceptual differences between various schools of psychiatry, but also to problems of terminology and translation. Rümke commented that the Dutch definition of schizophrenia was narrower than that of the Swiss; what Rümke called an atypical manic-depressive psychosis, the Swiss labeled schizophrenia.[63] The psychiatrist Pichot and his collaborators described how they had tried to set up an International Dictionary of Psychiatric Terms in 1950, and found that the project was nearly impossible. They had trouble translating the single French term, *délire*, into the two English and German terms, respectively, delirium and delusion and *Delirium* and *Wahn*.[64] Others pointed out differences in diagnostic trends between different hospitals in the United States. For instance, private hospitals were less likely to diagnose schizophrenia than state hospitals, opting for the less stigmatized diagnosis of psychoneurosis.[65] Finally, even within wards in the same hospital there were different diagnostic patterns that followed changes in ward administration.[66]

Rümke, who attended the conference, did not explicitly discuss or defend the praecox feeling, but made a strong argument for the necessity of a nosology of mental illness founded on a biological or etiological basis. The nosology that he advocated could be reliably based on the form of disorder, not on its content, which, as he noted, could be quite variable and produce a number of different mental pictures.[67] Indeed, the praecox feeling had the advantage of revealing the universal form of genuine schizophrenia, a claim that Rümke had made two years earlier at the Second International Congress of Psychiatry in Zürich in 1957.[68] At this conference, Rümke confidently declared that the clinician's feeling in the presence of the patient could often be more accurate than a description of his symptoms. Although he valued descriptive methods as the root of all science, and stressed the importance of description in the Kraepelinian tradition, he conceded that psychiatric descriptions were still rather crude:

> Our clumsy words are nearly always beside the mark. If we are to make progress, we have to approach the subject with a new open–mindedness, as if we observed the mentally deranged, in this case the schizophrenes, for the very first time… .Even if we unable to describe accurately what we observe, all the same we do observe something! Not infrequently we base a diagnosis on something we observe positively, but cannot communicate in words to others. We cannot 'word' or 'verbalise' it.[69]

Previous attempts at description, whether Krapelinian or Bleulerian, had their deficits. For instance, schizophrenic symptoms manifested in other disorders, and the Bleulerian depiction of the schizophrenic splitting of the personality was simply not accurate, according to Rümke. Despite these descriptive problems, however, the clinician had a very good sense of the referent of the term 'schizophrenic' and could identify cases of 'genuine schizophrenia,' even without being able to articulate it clearly.

That the praecox feeling was beyond words was therefore an advantage: it skirted these descriptive difficulties and formed a definitive guide to diagnosis. Rümke declared: 'If this feeling is not roused in me—in that case the above-mentioned criteria always lack the 'very definite' or 'schizophrenic' character—I resist the diagnosis of 'genuine schizophrenia' as long as I can.'[70] Genuine schizophrenia could be identified by its form, and not its content, and these formal aspects were recognizable universally, as Rümke declared, from Paris to Toronto, and often 'at a glance'. [71] The praecox feeling, then, at least in Rümke's formulation, could identify the form of the disorder, and thus was a direct guide to the diagnosis of genuine schizophrenia.

Although Hempel argued against the praecox feeling as a basis for constructing a psychiatric science, there were few publicly agreed upon standards to structure diagnosis at the time, and the feelings, experiences and sensitivity of the psychiatrist still functioned as important elements of diagnosis. Indeed, despite attempts to minimize the personal equation in psychiatric diagnosis, clinicians continued to report that some form of praecox feeling, often defined as a lack of rapport with a patient, was a reliable indicator of schizophrenia. The results of a 1962 study questioning clinicians regarding the presence and reliability of the praecox feeling in Germany found that 86 percent of respondents were familiar with the feeling, 54 percent said it was a reliable guide to diagnosis and 25 percent said it was a more reliable indicator than other symptoms.[72] In this last group were psychiatrists with extensive clinical experience, and those working in asylums and private practice, in distinction to clinics. In a 1973 WHO report for a flexible system for the diagnosis of schizophrenia, poor rapport as evident in the present state examination was one of the most reliable of symptoms, among the 12 identified.[73]

Contemporary psychiatrists who have an interest in phenomenological methods have reported that although the recognition of the lack of rapport with a schizophrenic patient or an autistic tendency is not easily operationalized, it still plays a fundamental role in diagnosis.[74] Others have taken issue with the DSM as a diagnostic tool precisely because it accords little place for the importance of such phenomena such as clinical judgment, clinical experience, countertransference and the praecox feeling.[75] These tacit elements of the doctor–patient relationship have been particularly resistant to evaluation and systematization. The Heidelberg psychiatrist Alfred Kraus has argued that a typological or holistic evaluation remains an important aspect of practice, even if it is neither explicitly grounded, nor accounted for in current classification schemes.[76]

For the phenomenologically oriented psychiatrists of the early twentieth century, diagnosing schizophrenia on the basis of feeling and intuition was closely linked to the moments of doctor–patient encounter. They borrowed an everyday, tacit knowledge of others, and honed it to formulate an embodied epistemology, or self-technology for medical diagnosis. This method led to an ontology of schizophrenia that saw its essence as an inability to connect with others and with one's world. But beyond defining the essence of schizophrenic pathology, this epistemology extended to an ontology of mind. Invoking sympathy, empathy, inner perception and intuition as varied, but related means for the direct contact with other minds, these psychiatrists embraced a view of mind not so much as a container holding various mental contents, but rather as extending beyond individual selves. The emphasis these clinicians placed on empathic and sympathetic connections to others, along with their pathological absence, may prompt us to consider more carefully Minkowski's paradox that 'our psyche is everywhere but in ourselves'.[77]

Notes

1. I gratefully acknowledge the helpful advice of G.E. Berrios, Louis Sass, Jacob A. Belzen and the editor of this volume, Fay Bound Alberti.
2. Michael Polanyi, *Personal Knowledge: Towards a Post-Critical Philosophy* (Chicago: University of Chicago Press, 1958), p. 263.
3. The cultural and medical historian Sander Gilman has noted that the term 'bizarre' has served as a technical term in twentieth century psychiatry. See Sander Gilman, 'Seeing the Schizophrenic: On the 'Bizarre' in Psychiatry and Art', *Disease and Representation: Images of Illness from Madness to AIDS* (Ithaca: Cornell University Press, 1988), pp. 231–44.
4. Eugen Bleuler, *Dementia Praecox oder Gruppe der Schizophrenien* (Tübingen: Ed. Discord, 1988 [1911]), p. 6.
5. Bleuler, *Dementia Praecox*, p. 262.
6. See for instance, G.J. Barker-Benfield, *The Culture of Sensibility: Sex and Society in Eighteenth-Century Britain* (Chicago: University of Chicago Press, 1992); Catherine Lutz, *Unnatural Emotions: Everyday Sentiments on a Micronesian Atoll & their Challenge to Western Theory* (Chicago: University of Chicago Press, 1988); Jan Goldstein, *The Post-Revolutionary Self: Politics and Psyche in France, 1750–1850* (Cambridge MA: Harvard University Press, 2005); Otniel Dror, 'The Affect of Experiment: The Turn to Emotions in Anglo-American Physiology, 1900–1940' *Isis* (1999), 90, pp. 205–37; and idem, 'Creating the Emotional Body: Confusion, Possibilities, Knowledge' in Peter N. Stearns and Jan Lewis (eds), *Emotional History of the United States* (New York: New York University Press, 1998), pp. 173–94; Joel Pfister and Nancy Schnog (eds), *Inventing the Psychological: Towards a Cultural History of Emotional Life in America* (New Haven: Yale University Press, 1997).
7. One recent exception is Riskin's discussion of scientific style and sensibility in the French enlightenment. See Jessica Riskin, *Science in the Age of Sensibility: The Sentimental Empiricists of the French Enlightenment* (Chicago: University of Chicago Press, 2002).
8. For more on phenomenologically inclined psychiatrists see Herbert Spiegelberg, *Phenomenology in Psychology and Psychiatry* (Evanston: Northwestern University Press, 1972); Torsten Passie, *Phänomenologisch–Anthropologische Psychiatirie und Psychologie: Eine Studie über den Wengener Kreis: Binswanger–Minkowski–von Gebsatte–Straus* (Hürtgenwald: Guido Pressler Verlag, 1995). Although these psychiatrists were familiar with psychoanalytic principles, they preferred their own experiential and emotional language in their assessments of both clinician and patient.
9. Although in recent years such empathic methods have been increasingly been understood as possessing a gendered or feminized inflection, in late nineteenth-century Central Europe this approach rather resonated with the image of the broad-minded, humanistic scholar. See Wilhelm Dilthey 'Ideas Concerning a Descriptive and Analytic Psychology' in idem, *Descriptive Psychology and Historical Understanding*, trans. Richard M. Zaner and Kenneth L. Heiges (The Hague: Martinus Nijhoff, 1977), pp. 41–81; Fritz Ringer, *The Decline of the German Mandarins* (Cambridge MA: Harvard University Press, 1969); Hans Herbert Kögler and K. R. Stueber (eds),

Empathy and Agency: The Problem of Understanding in the Human Sciences (Boulder: Westview Press, 2000). For more current work on gender and empathy see Ellen Singer More and Maureen A. Milligan (eds), *The Empathic Practitioner: Empathy, Gender and Medicine* (New Brunswick: Rutgers University Press, 1994).

10. Eugène Minkowski, *Le Temps Vécu: Études Phénomenologiques et Psychopathologuiques* (Paris: Collection de l'Évolution Psychiatrique, 1933) p. 167.
11. Carl Hempel, 'Fundamentals of Taxonomy' in *Aspects of Scientific Explanation and other Essays in the Philosophy of Science* (New York: The Free Press, 1965) pp. 137–54.
12. For a phenomenological account of the ways in which non-paranoid schizophrenics may experience affects, but not emotions, where emotions are defined as the basis for empathy and congruence with others, see Louis Sass, 'Affectivity in Schizophrenia: A Phenomenological View', *Journal of Consciousness Studies* (2004), 11, pp. 127–47.
13. Alfred Hoche, 'Die Bedeutung der Symptomenkomplexe in der Psychiatrie ' *Zeitschrift für die gesamte Neurologie und Psychiatrie* (1912), 12, pp. 540–51; Karl Bonhoeffer, 'Zur Frage der Exogenen Psychosen', *Zentralblatt fur Nervenheilkunde und Psychiatrie* (1909), 32, pp. 499–505.
14. See Leopold Bellak (ed.), *Contemporary European Psychiatry* (New York: Grove Press, 1961), pp.330, 343.
15. Wolfram Schmitt, 'Das Modell der Naturwissenschaft in der Psychiatrie im Übergang vom 19. zum 20. Jahrhundert', *Berichte zur Wissenschaftsgeschichte* (1983), 6, pp. 89–101. Nathan Hale speaks of the 'crisis of the somatic style', which peaked between 1904 and 1909 at least in American circles. Nathan G. Hale, Jr., *Freud and the Americans: The Beginnings of Psychoanalysis in the United States, 1876–1917* (New York: Oxford University Press, 1971). Emil Kraepelin was a convert to physiological psychology and his studies of physical and mental fatigue applied the concept of work to mental activity. See Anson Rabinbach, *The Human Motor: Energy, Fatigue, and the Origins of Modernity* (Berkeley: University of California Press, 1990), pp. 189–95.
16. The historian of psychiatry German E. Berrios has asserted that the labeling of the older descriptive method as phenomenological was a way of trying to shore up the decaying authority of the classic nineteenth-century descriptivism. See German E. Berrios, 'Phenomenology, Psychopathology and Jaspers: A Conceptual History', *History of Psychiatry* (1992), 3, pp. 303–27; and idem, 'Phenomenology and Psychopathology: Was There Ever a Relationship?' *Comprehensive Psychiatry* (1993), 34, pp. 213–20.
17. Karl Jaspers, *Allgemeine Psychopathologie* (Berlin: Julius Springer, 1913), translated as *General Psychopathology* (Chicago: University of Chicago Press, 1968), pp. 55–6. Some of these comments were added in a later edition. Although there has been a recent surge in understanding illness from patients' own accounts and memoirs, the phenomenological psychiatric tradition has emphasized the benefits of such an approach as a means for the scientific understanding of mental disorders.
18. For Jung, complexes were affectively motivated responses made manifest in reaction times to association tests. See Eugen Bleuler, *Affektivität,*

Suggestibilität, Paranoia (Halle: Verlag von Carl Marhold, 1906); Carl G. Jung, *Über die Psychologie der Dementia Praecox* (Halle: Verlagsbuchhandlung Carl Marhold, 1907).

19. See Henri Ellenberger, *The Discovery of the Unconscious: The History and Evolution of Dynamic Psychiatry* (New York: Basic Books, 1970), pp. 286–88; Roland Kuhn, 'Eugen Bleuler's Concepts of Psychopathology', *History of Psychiatry* (2004), 15, 3, pp.361–66.
20. On Binswanger's Bellevue asylum, and the assessment of reflective and spontaneous utterances of patients see Susan Lanzoni, 'Diagnosing Modernity: Mania and Authenticity in the Existential Genre', *Configurations* (2004), 12, pp. 107–31; eadem, 'Bridging Phenomenology and the Clinic: Ludwig Binswanger's Science of Subjectivity' (Unpublished Ph.D., Harvard University, 2001). For an account of patient reflections on their experiences of polio, and the predominance of such reports from middle class patients see Daniel J. Wilson, *Living with Polio: The Epidemic and its Survivors* (Chicago: University of Chicago Press, 2005).
21. Henricus Cornelius Rümke, *Problems in the Field of Neurosis and Psychotherapy* (Oxford: Blackwell Scientific Publications, 1953), p. 9. For a description of Rümke's eclectic approach that centered on the clinic see Jacob A. Belzen, 'The Impact of phenomenology on clinical psychiatry: Rümke's position between Jaspers and Kraepelin', *History of Psychiatry* (1995), 6, pp. 349–85.
22. Jaspers distinguished between 'Einfühlbares und Nichteinfühlbares (Natürlichenes und Schizophrenes) Seelenleben', *Allgemeine Psychopathologie* p. 89. He ascribed to Bleuler's definition of schizophrenia as accurately conveying its unnatural quality, p. 90.
23. Ludwig Binswanger, 'Welche Aufgaben ergeben sich für die Psychiatrie aus den Fortschritten der neueren Psychologie?' *Zeitschrift für die gesamte Neurologie und Psychiatrie* (1924), 91, pp. 402–36.
24. Bleuler, *Affektivität,* p. 12. Bleuler referred to the work of the psychologist Joseph Nahlowksy, who viewed feelings as closely connected to ideas. See Joseph Nawlowksy, *Das Gefühlsleben, in seinen Wesentlichsten Erscheinungen und Bezügen* (Leipzig: Verlag von Veit, 1884).
25. 'Wir diagnostizieren in unserm Fall eigentlich nicht nach, sondern *mit* dem Gefühl, d.h. mittels jener Wahrnehmungsweise, die mit dem 'Gefühl' im Sinne von sinnlichen oder emotionellen Gefühlen außer dem Namen nicht das geringste gemein hat . Binswanger, 'Welche Aufgaben', p. 427.
26. 'Ein Schizophrener kann mir sehr sympathische sein als Mensch, und trotzdem pralle ich innerlich zurück, erlebe ich immer wieder eine Schranke meiner inneren Einigung mit ihm, bei der Wahrnehmung seiner Person als einer schizophrenen', ibid.
27. See *Spiegelberg, Phenomenology in Psychology,* p. 237; Annick Urfer, 'Phenomenology and Psychopathology of Schizophrenia: The Views of Eugene Minkowksi' *Philosophy, Psychiatry and Psychology* (2001), 8, pp. 279–89.
28. Minkowski to Binswanger, 2 September 1924 and Minkowski to Binswanger, 15 March 1926, Universitätsarchiv Tübingen (UAT), University of Tübingen, Tübingen Germany, 443/24. Binswanger to Minkowski, 18 March 1926, UAT 443/24. For a discussion of inner and outer perception see Max Scheler, *Zur Phänomenologie und Theorie der Sympathiegefühle und von Liebe und Haß. Mit einem Anhang über den Grund zur Annahme des fremden Ich* (Halle: Niemayer,

1913), translated as *The Nature of Sympathy* (Archon Books: 1970). For more on Scheler's influence on Binswanger see Susan Lanzoni, 'An Epistemology of the Clinic: Ludwig Binswanger's Phenomenology of the Other', *Critical Inquiry* (2003), 30, pp. 160–86.
29. Eugene Minkowski, *La Schizophrénie* (Paris: Desclée de Brouwer, 1958 [1927]), p. 58.
30. Ibid., p. 57. Minkowski returned to this point in *Le Temps Vécu*, p. 65.
31. Minkowski, *Le Temps Vécu*, pp. 65–6.
32. Minkowski, *La Schizophrénie*, p. 57.
33. Minkowski's presentation 'Etude psychologique et analyse phénoménologique d'un cas de mélancolie schizophrénique' is reprinted in *Le Temps Vécu*, pp. 169–81; Ludwig Binswanger, 'Über Phänomenologie', *Zeitschrift für die gesamte Neurologie und Psychiatrie* (1923), 82, pp. 10–45; Eugen Bleuler, 'Korreferat', Schweizerischer Verein für Psychiatrie Protokoll der 63. Versammlung, *Schweizer Archiv für Neurologie und Psychiatrie*, (1923), 12, pp. 330–1.
34. Minkowski, *Le Temps Vécu*, pp. 170–1. trans. and adapted from Eugène Minkowski, *Lived Time: Phenomenological and Psychopathological Studies* (Evanston: Northwestern University Press, 1970), p. 182.
35. The section of his book is entitled: '2. –Nos propres reactions en presence du malade en tant que moyen d'investigation des troubles mentaux', Minkowski, *Le Temps Vécu*, p. 163.
36. Ibid., p. 165.
37. Ibid.
38. Ibid., p. 166.
39. Bleuler called autism one of the complexes, but basic symptoms of schizophrenia, defined as a detachment from reality coupled with the predominance of the inner life. Bleuler, *Dementia Praecox*, p. 52.
40. This had been the topic of his medical thesis: Eugène Minkowski, *La notion de perte de contact vital avec la réalité et ses applications en psychopathologie* (Paris: imprimerie de la Faculté de Médicine; Jouve & Cie. Éditeurs, 1926).
41. Minkowski, *La Schizophrenie*, p. 50.
42. Ibid.
43. Minkowski cited the work of Binet and Simon as having similar aims: 'Définition des Principaux états d'aliénation', *Année Psychologique* (1910), 16 as cited in Minkowski, *Schizophrenie*, p. 49. According to Minkowski, this synthetic method, born of long experience and the method of participation was not to replace a diagnosis based on the rational enumeration of symptoms and signs, but was a critical addition to it.
44. This was at a 1942 meeting of the Psychological Society in Bern, where Binswanger gave a lecture on the knowledge of human existence, entitled: 'Über Daseinserkenntnis'. In the discussion that ensued, at the suggestion of the psychiatrist Jacob Wyrsch, Binswanger's approach was renamed existential analysis (Daseinsanalyse). Gerhard Fichtner (ed.), *Sigmund Freud Ludwig Binswanger Briefwechsel 1908–1938* (Frankfurt am Main: S. Fischer Verlag, 1992), p. XXV. See Jakob Wyrsch, *Die Person des Schizophrenen* (Bern: Haupt, 1949).
45. Jakob Wyrsch, 'Über die Intuition bei der Erkennung des Schizophrenen ' *Schweizerische Medizinische Wochenschrift* (1946), 76, pp. 1173–6.
46. Jakob Wyrsch, 'Über die Intuition', p. 1173.
47. Ibid., p. 1175.

48. Wyrsch claimed that one could sometimes empathize with a schizophrenic, and conversely there were those with whom one could not emotionally identify (nachfühlen), but who were not schizophrenics but psychopaths. Wyrsch, 'Über die Intuition' p. 1176. Wyrsch was in accord with the work of the German psychiatrist Hans Grühle, who had also questioned Jaspers's focus on empathy. At mid-century there were also attempts to treat schizophrenics with psychotherapeutic methods, thus providing a model for the possibility of a different kind of rapport with the schizophrenic. See M. Sechehaye, *La realization symbolique* (Bern: Huber, 1947); and a series of papers presented on the topic in 'Symposium international sur la psychothérapie de la schizophrénie', in *Acta Psychotherapeutica* (1957), 5, pp. 2–4.
49. Reprinted and translated as 'The Nuclear Symptom of Schizophrenia and the Praecoxfeeling', *History of Psychiatry* (1990), 1, pp. 331–41. See also Rümke, 'Die klinische Differenzierung innerhalb der Gruppe der Schizophrenien', *Der Nervenarzt* (1958), 29, pp. 49–53, translated as 'The Clinical Differentiation within the Group of the Schizophrenias', in Werner A. Stoll (ed.), *Proceedings of the Second International Congress for Psychiatry* (Zürich: Füssli, 1959), vol. 1, pp. 302–10.
50. Rümke 'The Nuclear Symptom', p. 336.
51. Ibid.
52. Ibid.
53. Ibid., p. 337.
54. Rümke, Problems in the Field, p. 69.
55. Alice and Michael Bálint argued for increased control over the analyst's emotions in an important paper, 'On Transference and Counter-Transference', *The International Journal of Psycho-Analysis* (1939), 20, pp. 223–30. On the contributions of Ferenczi and Balint to the phenomenon of counter-transference see André E. Haynal, *The Technique at Issue: Controversies in Psychoanalysis: From Freud and Ferenczi to Michael Balint* (London: Karnac Books, 1988), pp. 19–34, 71–102.
56. Rümke, Problems in the Field, p. 72.
57. Joseph Zubin (ed.), *Field Studies in the Mental Disorders* (New York: Grune & Stratton, 1961) Funds to support the conference were obtained from the National Institute of Mental Health, and the conference was organized by Joseph Zubin, a Professor of Psychology at Columbia University. Zubin studied biometrics and epidemiology, and had undertaken a tour of European research centers in 1957.
58. E. Stengel, 'Afternoon Session: Problems of Nosology and Nomenclature in the Mental Disorders' in Zubin (ed.), *Field Studies*, p. 54.
59. Joachim-Ernst Meyer, 'Afternoon Session: Problems of Nosology and Nomenclature in the Mental Disorders' in Zubin (ed.), *Field Studies*, p. 66.
60. Carl Hempel, 'Fundamentals of Taxonomy' in *Aspects of Scientific Explanation and other Essays in the Philosophy of Science* (New York: The Free Press, 1965), pp 137–54, 142.
61. At the same time, however, Hempel indicated that operationism had its limits, and that it could not produce a full definition of a scientific term. For instance, temperature could be defined operationally not only by a mercury thermometer but also by an alcohol thermometer, and each operation only

provided a partial definition, or partial criterion of application for the scientific term 'temperature'. Hempel, 'Fundamentals of taxonomy', p. 143.
62. Ibid., p. 142.
63. Stengel, 'Problems of nosology', p. 59.
64. Comments by Dr. Pichot, 'Open Discussion' in Zubin (ed.), *Field Studies*, p. 98.
65. Morton Kramer, 'Open Discussion ', *Field Studies*, p. 115.
66. Benjamin Pasamnick, 'Open Discussion', *Field Studies*, p. 118.
67. Rümke, 'Problems of Nosology' in Zubin (ed.), *Field Studies*, p. 74.
68. Rümke, 'Clinical Differentiation', p. 305.
69. Rümke argued that it was not a true feeling, but an experience, which served as a compass to diagnosis. Ibid., p. 303.
70. Ibid., p. 305.
71. Ibid.
72. Gerhard Irle, 'Das praecoxgefühl in der diagnostik der schizophrenie: Ergebnisse einer Umfrage bei westdeutschen Psychiatern', *Archiv für Psychiatrie und Zeitschrift* für die gesamte Neurologie (1962), 203, pp. 385–406.
73. Only 'thoughts aloud' had a higher reliability (.95), than that of poor rapport (.86). William T. Carpenter, Jr. John S. Strauss and John J. Bartko, 'Flexible System for the Diagnosis of Schizophrenia: Report from the WHO International Pilot Study of Schizophrenia', *Science* (1973), 182, pp. 1275–8.
74. Josef Parnas and Pierre Bovet, 'Autism in Schizophrenia Revisited', *Comprehensive Psychiatry* (1991), 32, pp. 7–21. They note that in the DSM-IIIR the schizotypal personality is marked by restricted or inappropriate rapport, and social isolation (p.18). Psychiatric trainees have reported that experienced clinicians have advised them to document their own emotional responses to patients as a guide to diagnosis.
75. Manfred Spitzer, 'The Basis of Psychiatric Diagnosis', in John Z. Sadler, Osborne P. Wiggins, and Michael A. Schwartz (eds), *Philosophical Perspectives on Psychiatric Diagnostic Classification* (Baltimore: Johns Hopkins, 1994), p. 174.
76. Alfred Kraus, 'Phenomenological and Criteriological Diagnosis: Different or Complementary?' in Sadler, Osborne, Wiggins, and Schwartz (eds), *Philosophical Perspectives*, pp. 148–60. Other critiques of the DSM have argued that instead of moving toward greater theoretical sophistication as outlined by Hempel, the proliferation of disorders and syndromes has produced little scientific progress. William C. Follette and Arthur C. Houts, 'Models of Scientific Progress and the Role of Theory in Taxonomy Development A Case Study of the DSM', *Journal of Consulting and Clinical Psychology* (1996), 64, pp. 1120–32.
77. Minkowski, *Le Temps Vécu*, p. 62.

Index

"n" stands for a keyword present in a Endnote and "nn" stands for a term present in many notes

Abjection, 90–1
Affection (*See also* Emotion definitions), 23–31, 43–6, 50 n.66
  and 'passion', 31–5
  and pathologies, 31–5
Anaesthetic, 92, 101, 108, 112–14, 118
Anger, 3–6, 13–15, 54
Anthropology of emotion, xxvii n.25
Anti-Vivisectionism, 108, 111, 114, 116, 118
Anxiety, medical dangers of, 155
Aristotle, xviii
Asylum, 171

Balzac, Honoré D., 85–6, 97 nn.30, 31, 33, 35
Beebe-Center, John Gilbert, 126, 138 n.5
Bell, Sir Charles, 100, 111, 127, 139 n.9
Bentham, Jeremy, 38–9, 44, 51 n.70, 52 n.94
Bernard, Claude, 127, 134, 136, 142 nn.45, 47
Biology (*See also* Gender), 64, 73
Body, clinical and laboratory representations, 136
Boerhaave, Hermann, 2, 8–9, 11–12, 20 n.56
Bourdieu, Pierre (*See also Habitus*), xviii–xix, xxiii, xxviii nn.37, 40, 42
Breast-feeding and emotions, 34
British Medical Association, 106
Brown, Thomas, 26, 28–33, 37–8, 45, 47 n.16, 48 nn.29, 32, 33, 38, 49 n.48, 50 n.53, 52 n.97
Brutal passions, 106–9
Buchan, William, 53, 74 n.3
Bucknill, John Charles, 60, 72, 76 nn.31, 34, 78 n.78
Burdon Sanderson, John, 104–7, 113, 117, 120 nn.14, 15, 121 n.19
Burke, Edmund, 35–6, 50 n.61
Burton, Isabelle, 116, 123 n.58
Burton, Robert, 4, 18 nn.19, 23

Cambridge School of Physiology, 123 n.64
Cannon, W.B., 127, 130, 139 n.18, 140 nn.19, 26, 141 nn.28, 30, 39, 142 n.53
Cardiograph, 129, 134, 142 n.45
Carpenter, William B., 31, 34, 40, 48 n.38, 50 n.55, 51 n.76, 102–3, 111, 115, 120 n.4
Cartesian philosophy (*See also* Descartes), 7, 10
Catholicism, 85, 89–90, 92
Character (*See also* Morals; Influence), 39, 66, 70
Charlotte, Princess (*See also* Childbirth), 53, 74 n.6
Cheyne, George, 8, 11, 19 n.52
Childbirth, 53–65, 72–3, 75 nn.13, 26
Civilizing Process, The (*See also* Elias, Norbert), xxvii n.21, 1, 16 n.3, 81
Class and emotion, 64, 66, 68–9
Cobbe, Francis Power, 107–10, 115, 121 nn.23, 25, 27, 28, 32, 123 n.57
Cogan, Thomas, 44, 52 n.93
Communities of emotion, 56, 67–73
Compassion
  'compassionate conservatism', 80
  and empathy and humanity, 82–6
  and humanitarian protocols, 79–99
  and pain and religion, 89–93
Confinement (and pregnancy), 55, 57, 61–2, 66, 68, 70
Conolly, John, 69, 77 n.66

Corporeal, 4–5, 9, 14 (*See also* body, clinical and laboratory representations)
  and incorporeal functions, 8
  from mental to corporeal causation, 14
  in Descartes model, 9
Crile, George W., 132, 137, 141 nn.31, 33
Cullen, William, 8, 19 n.38, 30–2, 34, 39, 41, 48 n.38, 49 n.39, 50 n.53, 51 n.73

Darwin, Charles (*See also Expression of Emotion*), xx, xxi, xxviii n.44, 33, 35, 45–6, 48 n.33, 79, 95 n.5, 103, 110, 115, 119, 120 n.8, 123 nn.56, 69, 124 n.69, 127–8, 136, 139 n.12, 143 n.56
Darwin, Erasmus, 29, 33
Darwin, Francis, 110, 122 n.36
Daseinsanalyse method, 174
De Cyon, Elie, 108, 121 n.28, 123 n.57, 127
Delusional schizophrenic patient, 176
Descartes, René (*See also* Reflex), 2, 8–9, 11, 18 n.18, 19 nn.39, 42, 127, 139 n.7
Diderot, Denis (*See also Letter on the Blind*), 83, 85, 96 n.21
Domesticity (*See also* Motherhood), 70
Duchenne de Boulogne, 127–8
Dunant, Henry (*See also Memory of Solferino, A*), 82, 91–5, 96 n.16, 97 nn.37, 38, 39, 41, 98 nn.44, 45, 99 nn.62, 71
  and humanitarian narratives, 86–9

Edinburgh, physicians and metaphysicians in, 27–31
Electrical impulses, 1
Elias, Norbert (*See also Civilizing Process*), xvi, xxvii n.21, xxviii n.40, 1, 16 n.2, 68, 77 n.63
Emotionology (*See also* Stearns, Peter N.), xvii, xxvi n.17, xxvii n.29
Emotions
  'Emotional Communities' (*See also* Rosenwein, Barbara), xvii, 68–79
  and experimentation, in late Victorian medicine, 100–24, (*See also under* experimentation)
  and patients and passions, 22–52 (*See also separate entry)*
  and puerperal insanity medicalization, 100–24
  as mechanistic components of body-as-machine, 126
  changing models of, 159
  definitions (*See also* Passions, Affection), 24–7, 28–31, 34, 44–6, 129–30
  historical unconscious, emergence, 145–61
  historiography, xv–xvi, xx, xxvi n.20
  humouralism, classical inheritance of, 2–7
  in British medical practice, 144–68
  in illness, 153
  in the early modern medical tradition, 1–21 (*See also separate entry)*
  instrument-generated graphic representations of, 134
  languages and landscapes of, 53–78 (*See also under* Motherhood)
  materialization and mechanization of, 128
  mechanical philosophy and hydrodynamic principles, 7–12
  modern study, 127
  nerves, role of, 12
  pathologization, 53–9
  patient–physician relation in, 154–8, 184
  producing, 130–4
  somatic expressions of, 137
  theory, and medical history, xiv–xxix
  visual representations of, 135

Empathy, xiv, xix, xxi–xxii, xxiv n.3, 81, 91–2, 95 nn.7, 8 and humanity, 82–6
Enlightenment, 1, 17 n.4, 18 n.19, 19 nn.48, 49, 80–5, 89–90
Ergograph (*See also* Mosso, Angelo), 126
Ethics (*See also* Morals), 37–9, 43, 50 n.63
European psychiatry, clinical experience, 171–3
Experimentation (*See also* Vivisection)
 and emotion in late Victorian medicine, 100–24
 brutal passions, 106–9
 feeling with instruments, 104–6
 reflexive physiology, 102–3
Expression of the Emotions (*See also* Darwin, Charles), xx, xxviii n.44, 45, 79, 103, 110, 120 n.8, 122 n.36, 123 n.69, 124 n.69

Falconer, William, 31, 48 n.38, 49 n.40
Fatigue, 125–6
Fear (of childbirth), 57, 59–60, 66–7, 73, 75 n.13
Fear (*See also* Mosso, Angelo), 139 n.18
 and emotions, 129–34
 and loathing in the laboratory and clinic, 125–43
 representations, 134–6
Feeling-diagnosis, 169–90 (*See also* Emotions)
 as a kind of 'emotional perception', 174
 feeling with instruments, 104–6
Foster, Michael, 105, 110, 113, 122 n.47, 123 n.64
Foucault, Michel (*See also* Madness; Moral Treatment), 24, 43, 47 n.12, 52 n.89, 68, 77 n.64
Framing, xvii, 60–8
French Revolution, 24, 30, 46, 50 n.61, 82, 83
Freudian theory, 154

Galenism, and human body, 3
Gefühlsdiagnose, 174–5, 178
Gender, xv, xvi, xix, xxii
 and emotion, 4, 15
 and science, 115–16
Geneva Convention, 86
George III (*See also* Madness), 38, 46 nn.2, 6, 7, 47 n.10, 51 n.78
Gestation, 59
Gooch, Robert, 59, 64–5, 69–73, 75 n.25, 76 nn.45, 47, 48, 77 n.54, 55, 68, 69, 78 nn.74, 79
Gregory, James (*See also* James, Gregory C.), 28–9, 31, 34, 37, 39–40, 48 n.38, 49 n.39, 49 n.46, 50 nn.63, 64, 51 n.69
Gregory, James C. (*See also* Gregory, James), 34, 41, 50 n.53,
Gregory, John, 30–1, 33, 36–8, 49 n.39, 49 n.46, 50 n.63, 50 n.63
Grief, 6, 14–15

Habitus (*See* Bourdieu, Pierre), xix, xxviii n.40
'Hallucinations of dread', 133
Harvey, William, 7, 10, 17 n.8, 18 nn.18, 33, 19 n.51
Heart, and emotion, 2–7, 9, 13–16, 18 n.26
Hippocrates/Hippocratic, 2, 14
Hobbes, Thomas, 3, 18 n.15
Hoffman, Friedrich, 8, 11
Hope, 33
Humanitarian protocols, 79–99 (*See also under* Pity; Compassion)
Humanitarianism, 79–86, 94
Humouralism, classical inheritance of, 2–7
Hutton, R.H., 109–10, 112, 115
Huxley, Thomas, 103, 111–12, 114–16, 122 n.41
Hypochondria (*See also* Melancholy/Melancholia), 12, 15, 20 nn.65, 77
'Hypochondria epidemica', 157
Hysteria, electrophysiological investigation, 151

Iatromechanism, 1
Infanticide, 62, 69
Influence (*See also* Morals), 56, 59, 63, 65, 69
Irritability and emotion, 64

James, William, xx, xxi, xxviii nn.45, 46, , 26, 45, 47 n.16, 126–7, 129, 142 n.52, 143 n.58
Joy, 24, 32–4, 40, 44

Kiss-O-Meter, 136
Klein, Emmanuel, 105, 113–14, 116

Laboratory medicine, 2
Lange, Carl George, xx, xxviii n.46
Laqueur, Thomas, 80–1, 93, 96 nn.9, 10
Letter on the Blind (*See also* Diderot), 83
Locke, John, 10, 16, 19 n.51
Lombard, W.P., 127

Madness (*See also* Foucault; Phrenzy), 31, 34, 60, 63–5, 71
  and morals, 38–43
Marey, Étienne Jules (*See also* Sphygmograph), 104–5, 120 n.9
Martineau, Harriet, 84, 96 n.28, 97 n.29
Measuring emotion, 129–36
Mechanical philosophy and hydrodynamic principles, 7–12
Mechanism (*See also* Vitalism), 9, 126
Melancholy/Melancholia (*See also* Hypochondria), xv, xxv, 3–7, 11–12
Memory of Solferino, A (*See also* Dunant, Henry), 82, 86, 88, 96 n.16, 97 nn.37, 41
Menstruation, 15
Method of 'penetration', 176
Military imagery, 111
Modernity, 1
'Moral' management 43
Morals/Moral treatment (*See also* Madness), 24–38, 44–6, 51 n.75
  and madness, 38–43
Mosso, Angelo (*See also* Ergograph; Fear), 125–8, 133–4, 136, 138 n.1, 141 n.36, 143 nn.55, 56
Motherhood (*See also* Domesticity)
  and puerperal insanity in nineteenth century, 53–78
  as emotional communities, 67–73
  emergence, 53–9
  gestation, 59
  melancholia (*See also* Hypochondria), 55

'Negative' emotions, 6
Nerves/Nervousness, 12, 15, 58, 63–5, 72
Neuropsychology, 1
Neutral/Zero emotional state, 132, 141 n.38
Newton, Isaac, 1, 7–8, 19 n.34
Nobel Prize, 86
Non-naturals, 3, 15

Pain, 81–84, 88
  and compassion and religion, 89–93
Passions, pathological (*See also* Emotions), 25–6, 49 n.41, 52 n.93
  and 'affections', 31–5
  and emotions, 40–2 (*See also* Patients and passions)
  and patients, in eighteenth-century, 22–52
  and politics, 35–8
  management, 50 n.49
  theories of, 25–7
  to 'emotions', transition, 29
Pathologies of emotion, 31–5
Patients and passions, 22–52 (*See also under* Emotions)
Philosophy of Emotion, 13, 16, 17 n.6
  mechanical philosophy and hydrodynamic principles, 7–12
Philosophy, moral, 28–31, 46
Phrenzy (Temporary; *See also* Madness), 55
Physiology and Emotion, 4–5, 15, 30, 33–4, 40, 126–7, 136, 138, 140 n.26

Pitcairne, Archibald, 2, 7–8, 11, 13
15, 17 n.10
Pity
and compassion and humanitarian protocols, 79–99
compassion, pain and religion, 89–93
empathy and humanity, 82–6
Politics of Emotions, 35–8
'*Praecox* feeling', 169, 179–84
Pre-Cartesian model, 10
Psyche, xix, xx
and *soma*, 2, 5, 9, 16
Psychiatric taxonomy, problems of, 181–84
Psychoanalysis, of emotions, 144–68
Psychology, and Emotion, 2, 7–8, 13, 15, 17 n.10
Psychology, Victorian, 29–30
Puerperal insanity, 53–78 (*See also* Motherhood)
and household, 67–73
as female disorder, 58
emergence, 55
framing, 59–67
identification, 62
in asylum casebooks, 65–6
nature of, 72–3
symptoms, 70

Quasi-mesmeric technique, 22–3
Queer studies, 90

Reason and emotion, 24, 26, 34, 36–7, 40–4
Red Cross, 79–80, 86–7, 94, 98 n.45
Redemption, 89
Reflex (*See also* Descartes), 8–9, 102–3
Reid, Thomas, 32, 49 n.46
Religion (*See also* Soul), 27, 31, 37, 46, 51 n.66, 62, 89
Representational level, of emotions, 134
Rosenwein, Barbara H. (*See also* 'Emotional Communities'), xvi–xviii, xxvii nn.23, 31, xxviii n.36
RSPCA (Royal Society for the Prevention of Cruelty to Animals), 106–8, 110, 112
Rush, Benjamin, 31, 34–6, 42, 44, 48 n.38, 50 nn.53, 54, 57, 62, 51 nn.84, 85, 52 n.96

Schizophrenia
clinical assessment, in early twentieth-century European psychiatry, 169–90
definition, 172
diagnosing with affect, 173–81
genuine schizophrenia, 183
phenomenological methods, 184
Scott Sherrington, Sir Charles, 127, 134
Sensibility, 63–5
Sentimental vivisector and the animal automaton, 109–14
Sexual passion, 40
'Sham fight[s]', 132–3
Slavery, 81, 87–8
Sociology of medicine, xiv, xxvi n.19
Soma, xix, 2–5, 9, 14, 16
Somatically oriented medicine, 10
Sorrow, 4, 6, 14
Soul, role in emotions (*See also* Religion), 8–12, 15–16, 21 n.83
and heart, 2–7
Sphygmograph (*See also* Marey, Étienne Jules)
Spirits, Animal, 6, 8–9, 11, 16, 18 n.31, 26, 37
Spirits, Vital, 9–10
Stearns, Carol Z. (*See also* Emotionology), xv, xvii, xxvi n.17
Stearns, Peter N. (*See also* Emotionology), xv, xvii, xxvi nn.17–18, xxvii n. 29
Sufferings, mediating, 86–9
Sympathy, 32, 37, 79, 82–3, 85, 95 n.7

'Temporary phrenzy', 55
Therapeutics, 46
Tuke, Daniel Hack, 60, 72, 76 nn.31, 34, 78 n.78

Tuke, John Batty, 67, 77 n.58

Unconscious
emergence, 145–61
reification, 146

Vitalism (*See also* Mechanism), 10
Vivisection (*See also* Experimentation; Anti-Vivisectionism), 100–1, 105, 107–18, 119 n.2, 122 nn.38, 48
Von Haller, Albrecht (Albertus), 2, 8, 14–15, 20 nn.60, 61, 66, 71, 74, 138 n.6

Whytt, Robert, 13, 20 n.65
Will, 102–3, 107, 110, 114
Willich, Anthony Florian Madinger, 31–4, 44, 49 n.47, 50 n.50, 52 n.95
Willis, Francis, 22–4, 38, 42
Willis, Thomas, 8
Wright, Thomas, 3–5, 9, 18 nn.16, 17, 18, 21, 27, 19 n.48, 20 n.70